Lecture Notes in Mobility

The book series Lecture Notes in Mobility (LNMOB) reports on innovative, peer-reviewed research and developments in intelligent, connected and sustainable transportation systems of the future. It covers technological advances, research, developments and applications, as well as business models, management systems and policy implementation relating to: zero-emission, electric and energy-efficient vehicles; alternative and optimized powertrains; vehicle automation and cooperation; clean, user-centric and on-demand transport systems; shared mobility services and intermodal hubs; energy, data and communication infrastructure for transportation; and micromobility and soft urban modes, among other topics. The series gives a special emphasis to sustainable, seamless and inclusive transformation strategies and covers both traditional and any new transportation modes for passengers and goods. Cutting-edge findings from public research funding programs in Europe, America and Asia do represent an important source of content for this series. PhD thesis of exceptional value may also be considered for publication. Supervised by a scientific advisory board of world-leading scholars and professionals, the Lecture Notes in Mobility are intended to offer an authoritative and comprehensive source of information on the latest transportation technology and mobility trends to an audience of researchers, practitioners, policymakers, and advanced-level students, and a multidisciplinary platform fostering the exchange of ideas and collaboration between the different groups.

Gereon Meyer · Sven Beiker
Editors

Road Vehicle Automation 10

Springer

Editors
Gereon Meyer
VDI/VDE Innovation + Technik GmbH
Berlin, Germany

Sven Beiker
Stanford University
Palo Alto, CA, USA

ISSN 2196-5544 ISSN 2196-5552 (electronic)
Lecture Notes in Mobility
ISBN 978-3-031-34756-6 ISBN 978-3-031-34757-3 (eBook)
https://doi.org/10.1007/978-3-031-34757-3

This Springer imprint is published by the registered company Springer Nature Switzerland AG
The registered company address is: Gewerbestrasse 11, 6330 Cham, Switzerland

Preface

As we are editing the 10th volume of the Road Vehicle Automation books, we become aware of the tremendous progress that the field of automated mobility has made over the last decade. Building on the progress in advanced driver assistance systems and vehicle-to-vehicle communication, automated driving promised a safer, cleaner and more convenient transportation of people and goods, and an utopia of self-driving pods seemed near to reach. In the meanwhile, many lessons had to be learned, and a Gordian knot of technical, societal, economic, human and legal factors was untied for letting highly automated driving functions and first regular autonomous services like robotic taxi and delivery hit the road. And, still there is more to come, once the opportunities of big data and artificial intelligence are fully exploited and the innovation cycles for automated vehicles are accelerated by fully decoupled software and hardware control architectures.

In that sense, it continues to hold true that this compendium is a good indicator for the state of the industry and with that for how close automated road vehicles are to wide deployment. We are glad to publish the latest on user studies, business models, technology solutions, collaboration efforts, policies, and more with this collection of contributions from the Automated Road Transportation Symposium 2022 (ARTS22) that took place in Garden Grove, CA, USA in July 18–21, 2022.

We wish to thank all contributors to this book. Not only did they prepare and present their talks or breakouts at the Symposium, they also spent the extra time to write the chapters that make Road Vehicle Automation 10 the essential book that it is. Our special thanks go to Jane Lappin, Valerie Shuman, and Steven Shladover for organizing with their team the Automated Road Transportation Symposium in the first place, and for helping with the creation of this book, including their introduction that provides an insightful and concise overview of the Symposium. We would also like to thank the teams at Springer Nature and VDI/VDE-IT, here particularly Meike Brandt, for taking care of so many editorial and administrative tasks that are crucial for turning manuscripts into the final publication.

We hope that all readers will find valuable insights in the book, maybe this will bring back great memories from ARTS22, and we hope to see everybody again at ARTS23 in San Francisco, CA, USA.

April 2023

Gereon Meyer
Sven Beiker

Contents

Introduction: The Automated Road Transportation Symposium 2022

Steven E. Shladover[1](✉), Jane Lappin[2], and Valerie Shuman[3]

[1] University of California PATH Program, 1357 South 46th Street, Building 452, Richmond, CA 94804, USA
steve@path.berkeley.edu
[2] TRB Vehicle-Highway Automation Committee, Belmont, MA, USA
[3] Shuman Consulting Group, LLC, TRB Challenges and Opportunities of Road Vehicle Automation Subcommittee, Skokie, IL, USA
vs@shumangroupllc.com

Abstract. In 2022, the Automated Road Transportation Symposium returned to a traditional face-to-face meeting format after two years of virtual meetings caused by the global pandemic. The plenary presentations and breakout discussions continued to provide the meeting participants with the most up-to-date and authoritative information about the current international state of development and deployment of road vehicle automation systems, retaining its standing as the essential global meeting for industry, government and research practitioners in the field.

Keywords: Road vehicle automation · Road transport automation · Automated vehicles · Autonomous vehicles · Self-driving vehicles

1 Overview

The 2022 Automated Road Transportation Symposium (ARTS22) was organized and produced by a large team of professional volunteers working under the auspices of the National Academies of Science, Engineering and Medicine (NASEM) Transportation Research Board (TRB). The meeting was organized to serve the participants' interests in understanding the impacts, benefits, challenges and risks associated with increasingly automated road vehicles and the environments in which they operate. It brought together key government, industry and academic experts from around the world with the goal of identifying opportunities and challenges and advancing Automated Driving System (ADS) research across a range of disciplines.

The symposium was held at the Hyatt Regency Orange County Hotel in Garden Grove, CA from 18–21 July 2022. The plenary sessions were scheduled for the full mornings of the second and fourth days and half of the morning on the third day. The afternoons of the first three days were devoted to full-length breakout sessions, and half-length breakout sessions were held on the morning of the third day. Five parallel breakout sessions were held in each of the breakout time slots, for a total of twenty breakout sessions.

G. Meyer and S. Beiker (Eds.): ARTSymposium 2022, LNMOB, pp. 1–9, 2023.
https://doi.org/10.1007/978-3-031-34757-3_1

The breakout sessions were organized by committees of volunteers to address a wide range of topics. These were clustered into three thematic tracks to make it easier for attendees to identify the sessions of strongest interest to them:

- Policy
- Operations
- Safety.

The plenary and breakout session programs were planned and produced by the ARTS22 Planning Committee, which included a mixture of TRB volunteers and support staff from Noblis:

John Craig, Noblis; Richard Cunard, Engineer of Traffic and Operations, TRB; Raymond Gerte, Noblis; Cynthia Jones, Drive Ohio; Jane Lappin, TRB Vehicle-Highway Automation Committee Chair; Steven Shladover, University of California PATH Program (and former chair of the TRB Vehicle-Highway Automation Committee); Valerie Shuman, Shuman Consulting Group, LLC and Chair, TRB CORVA Subcommittee; Egan Smith, U.S. DOT Intelligent Transportation Systems Joint Program Office, and Edward Straub, SAE.

2 Keynote Talks

The plenary program began with pre-recorded welcoming remarks by Secretary of Transportation Pete Buttigieg, followed by a more detailed in-person presentation of the U.S. DOT perspective by Dr. Vinn White, Senior Advisor for Innovation in the Office of the Secretary of the U.S. DOT. Secretary Buttigieg identified a goal for automated vehicles to achieve safety better than human drivers, and emphasized the importance of seeking equity in the job opportunities that would be available for workers in the automated driving industry and industries affected by driving automation. Dr. White described automated driving as "how we move better" and emphasized the importance of safety in planning for widespread deployment.

Dr. Steven Cliff, the Administrator of the National Highway Traffic Safety Administration, gave the keynote talk to kick off the third day of the symposium. He noted that NHTSA is studying the crashes that have occurred involving automated driving systems to try to understand whether they were caused by potential defects, which could be cause for recalls. He advocated for a broad "safe systems approach" in automated driving system development, including careful consideration of equity issues related to who is likely to be injured in crashes during development and testing as well as deployment.

The final day of the Symposium began with a keynote talk by Dr. Robert Hampshire, the Deputy Assistant Secretary for Research and Technology and Chief Scientist of the U.S. DOT. He gave a comprehensive overview of the R&D activities that the U.S. DOT is sponsoring to accelerate progress toward deployment of automated driving systems across the various transportation modes, under the broad theme of "transformation". Dr. Hampshire reviewed the existing research programs as well as the relevant new DOT initiatives such as the Highly Automated Systems Safety Center of Excellence (HASS-COE) and the ARPA-I program to create more intelligent transportation infrastructure.

3 Plenary Panel Sessions

ARTS22 extended the trend from previous years of devoting a majority of the plenary program time to panel discussion sessions on important topics, featuring groups of speakers responding to questions from the moderator and interacting with each other, with only a limited number of formal presentations. These sessions also provided opportunities for audience members to submit questions through a web-based service. The panel discussion sessions covered a wide range of topics in policy, technology and applications of road transportation automation.

3.1 Real-World Automated Trucks: What It Takes to Integrate with Today's Fleet Operations

Richard Bishop organized and moderated this panel of speakers from Waymo and TuSimple and their respective fleet customers C.H. Robinson and Loadsmith. They discussed their ongoing field testing hauling freight for commercial customers using their automated driving systems under the supervision of test drivers. They agreed that the adoption of the technology would initially be focused on the larger carriers, and that it would be a gradual process, beginning with limited long-haul routes and gradually expanding to more routes. One of the main themes was the improvement of work-life balance for drivers, particularly with the shift of driving assignments from long-haul toward local drayage operations. The labor impact of the introduction of intermodal freight using trailers and containers on railroad flatcars for long-haul routes was cited as an analogy – creating a larger number of driving jobs in other parts of the logistics chain than the jobs that were eliminated in the long-haul operations.

3.2 Automated Transit Projects

Henriette Cornet from UITP introduced the large-scale European project SHOW, which is field testing automated transit vehicles in multiple cities in Europe, and offered that as a basis for discussion by representatives of transit automation projects that are being initiated in Jacksonville, Trenton and Connecticut. The Jacksonville and Connecticut projects simplify the technical challenges by operating the automated buses in dedicated facilities avoiding mixed traffic interactions, while the Trenton project aims at a more challenging application in mixed traffic but at low speeds. The developers of these projects explained some of the surprising challenges that they encountered as they worked toward implementation of their systems.

3.3 State and Local Government Approaches to Regulating Automated Driving

Steven Shladover organized and moderated this panel of speakers representing the states of Texas, Arizona and California and the City and County of San Francisco, providing diverse perspectives on regulating the testing and public deployment of automated driving systems. Together, these jurisdictions are hosting a large majority of the current automated driving mileage in the U.S. All of them expressed interest in protecting their citizens and visitors from unreasonable risks associated with automated driving, while encouraging automated driving innovations that are expected to produce long-term benefits in mobility and safety, but they adopted different approaches to achieving those goals.

3.4 Private Sector Perspectives on AV Public Policy

Ariel Wolf of Venable LLP organized and moderated this panel of representatives from Aurora, Embark, Waabi and Waymo discussing their perspectives on the key public policy issues that they face. The trucking automation applications face some additional regulatory concerns based on FMCSA requirements, but the NHTSA safety regulatory issues that apply to all vehicles are likely to be more complicated. There was general agreement that the regulatory approach will need to be phased, and there was broad interest in having NHTSA define a consistent national approach, with a single set of data reporting requirements, but they were also reluctant to share much data, especially if the data were to be made public. The panelists were receptive to having active federal regulations to facilitate public acceptance and protect the industry as a whole from potential "bad actors" working in this field.

3.5 Implementation of SAE Automated Vehicle Safety Consortium (AVSC) Best Practices

The Automated Vehicle Safety Consortium (AVSC) was established as one of SAE's Industry Technologies Consortia (ITC) to provide a mechanism for companies to cooperate on pre-competitive aspects of AV safety. Amy Chu, the Director of AVSC, moderated a panel discussion with representatives from member companies Ford, Volkswagen, Honda and Aurora. They discussed topics that they have worked on in AVSC, including how to allow passengers to interrupt an automated driverless trip for an emergency, what automation-relevant data to save in event data recorders for crash reconstruction, how to develop Safety Management Systems (SMS) and how ADS-dedicated vehicles should interact with vulnerable road users.

3.6 Discussion of Primary Technology Challenges to Widespread Deployment of Automated Driving Systems

Steven Shladover moderated this discussion with representatives of Apex.AI and Edge Case Research, companies that provide services to automated driving system developers. They discussed the primary technological issues that remain unresolved in developing verifiably safe automated driving that can work under a wide enough range of conditions to be commercially viable. The topics included software safety design, verification and validation; hazard perception and prediction; safety assurance for machine learning systems; scalability and portability of systems to new locations and vehicle platforms; high-fidelity simulation development and validation; and identification of sufficient scenarios to support robust safety cases.

3.7 Building a Win/Win: AV/Infrastructure Collaboration

Valerie Shuman moderated a discussion with representatives from Cavenue, MAPtm, The Eastern Transportation Coalition (TETC) and General Motors. A fundamental issue going forward is how AVs and the infrastructure can share both real-time and analytics data to make the whole transportation system safer and more efficient. The panel discussed how data can be used to enable AVs (e.g., by extending ODDs) and what we are doing in this area already, including solutions that augment sensor perception range and provide complementary/redundant sensor information; how data will be used to support IOO safety and efficiency goals and what we are doing in this area already, including using vehicle data to support data-driven decisions and to provide complementary/redundant sensor information; current challenges, such as determining appropriate data for AV use and expected outcomes, data quality, usability and trustworthiness, standards across geographies and vehicles, coverage, business models and data vs insights; and the need for practical collaborative research and strong stakeholder communication and buy-in.

3.8 The Last Word: An Informed Discussion with Veteran Industry Journalists

Jane Lappin organized and moderated this discussion with journalists from Bloomberg and Forbes who have extensive experience following developments in the automated driving industry. They emphasized the significance of the ongoing industry consolidations and the parallel dearth of new investment capital, forcing the industry to concentrate on generating near-term revenues. They agreed that package delivery is a more promising near-term target than automated ride-hailing, particularly given the current public reluctance to share rides. They thought that regulations will be needed on labeling or naming of systems in order to reduce the current level of consumer confusion about system capabilities. Looking forward, they saw the industry as fragile because of the combination of technological vulnerabilities and lack of profitable business models.

4 Plenary Presentations

Individual presentations were distributed across the plenary program in between the panel discussions to avoid Powerpoint fatigue from too long a sequence of consecutive presentations. Three of the presentations were given by speakers who were invited to cover specific topics that the planning committee believed to be important for the audience to learn about and the other three presentations were progress reports on some of the most important public-sector activities around the world related to automation (the U.S., European Commission and Japan).

4.1 Presentations on Specific Topics

- Laura Fraade-Blanar, Waymo – What is "Good" Driving? Framing Evaluation of Autonomous Driving Behavior through Drivership
- Raquel Urtasun, Waabi – Waabi's AI-first Approach to Scaling Self-driving Safely and Rapidly
- Chaiwoo Lee, MIT AgeLab – Public Knowledge of and Attitudes toward Vehicle Automation: Trends and Implications

4.2 National and International Government Activities Relevant to Automated Driving

- Earl Adams, Federal Motor Carrier Safety Administration – ADS Trucks – The FMCSA Perspective on Road Safety
- Andrea DeCandido, European Commission DG-RTD – The European Experience: A Structured Approach to Cooperative, Connected and Automated Mobility (CCAM)
- Yoichi Sugimoto, SIP-adus Program (Japan) – Towards Social Deployment of Automated Driving – SIP-adus Activity in Japan

5 Breakout Sessions

ARTS breakouts gather key experts from around the globe for more in-depth consideration of specific topic areas. The goal of the breakout sessions is to collaboratively answer the questions: *What needs to be true to make the AV vision become a reality? How can our research help drive progress year on year?* The 2022 program included 20 sessions and covered a wide range of specialized topics from across the field to enable this discussion for the industry as a whole (see program list below).

The primary findings from each afternoon's breakout discussions were reported back to the plenary the following day. The combined summaries provided in these Daily Roundups distill the latest insights from across the industry. The major focus for 2022, which cut across breakouts in all three tracks, was *integration.* In an evolution from prior years, which focused on collaboration, many discussions highlighted the need for integration in three dimensions:

- Geographic. Isolated local solutions are not the way forward. We need global, national, and regional approaches for all of the core AV industry building blocks, including strategies, roadmaps, frameworks, standards, terminology and definitions, and policies.
- System-level. The role of infrastructure as both a support for AV operation and a beneficiary of AV insights was highlighted this year. We need to leverage the strengths and address the weaknesses of human, ADS *and* infrastructure performance for a successful overall system. This need for three-way coordination came up in a wide range of topic areas, including trucking, teleoperation, traffic management, first responders, construction and inspection zones, rural use, climate change, transit, digital infrastructure, and cybersecurity.
- Data. The critical need for effective data-sharing continued to be a key area of discussion in 2022. Data types of interest included test data, live and historical road network operational data, asset and road environment data, monitoring data, and modeling data. Research, Planning, and Operations were all viewed as both producers and audiences for this data, while challenges around data capture, communications, and developing the open standards and specifications necessary to enable sharing continue to be important areas for ongoing work.

Breakout participants collaborated to develop new ideas, such as a CAV cybersecurity ecosystem map. Other sessions used Design Thinking and Wargames methodologies to spark high-energy discussions among the participating experts. Additional highlights included:

- The developers of safety assurance framework standards reported good progress.
- The NHTSA Partnership for Analytics Research in Traffic Safety (PARTS) has established a much-needed safety data sharing model, with initial crash analysis data expected in fall 2022.
- First responder groups are now coordinating to develop common use cases.
- The NIST OES has developed a structured description of the operating environment to support testing & certification.

5.1 ARTS22 Breakout Sessions

5.1.1 Policy Sessions

- Unscrambling the Automated Vehicle (AV) Policy Puzzle: AV Policy Development and Regulation Under a New Normal
- Mitigating Climate Change with ART Technologies
- Shark Tank: Is it Time for AVs to Grow Up?
- Beyond the DriveTrain: Achieving Efficiency in CAVs through Technologies and Regulations

5.1.2 Operations Sessions

- Automated Trucking Research and Development
- Digital Infrastructure for Roadway Transportation and Automation Integration
- Inconsistency of AV Traffic Flow Impacts: Predictions in Literature
- Automated Vehicle Technologies for Crowd-Sourced Roadway Environment Assessment
- Remote Assistance and Teleoperation for Automated Vehicle Operations
- AVs in Rural America: What Can We Learn from the Data?
- Enhancing Mobility with Automated Shuttles and Buses
- How Connected Vehicle Deployment Lessons Lay the Groundwork for Highly Automated Vehicles
- Evaluating First Responder Interactions as the AV Market Expands
- Interactive Traffic Management for Highly Automated Vehicles

5.1.3 Safety Sessions

- Safety Assurance of Automated Driving
- ADS Standards Hot Topics: Operational Design Domain (ODD) & Operating Envelope Specification (OES)
- Cybersecurity Hot Topics
- AV Testing and Data Collection
- Adapting War Games to Explore Safety Measurement: An Interactive Exercise
- Understanding the Human Factors of Teleoperation

6 General Cross-Cutting Observations

As the field of road vehicle automation has advanced and the level of knowledge of the issues has grown over the past several years, the areas of emphasis within the Symposium have continued to evolve. Based on the discussions at this most recent meeting, three broad categories of observations are worth noting:

6.1 Existing Trends that Accelerated Based on the COVID-19 Pandemic

- The urgency of identifying robust business models to achieve commercial success with automated driving products and services became more acute. New investments in automated driving businesses declined more dramatically than previously, as more investors became conscious of the development challenges and electrification became a more attractive market opportunity. Because of the difficulty of attracting additional investments, the existing companies experienced a more urgent need to generate revenue and shift attention to nearer-term and less technically ambitious market opportunities.
- The consolidations within the industry through mergers, acquisitions and corporate failures accelerated. This was related to growing recognition of the technological challenges to achieving high levels of automation and of identifying market niches that are both technologically and commercially feasible.
- The pandemic-inspired shift to work from home, online shopping and home delivery of fresh food and prepared meals accelerated the shift of interest from passenger movement to goods movement. The decline of commuting travel and the health concerns associated with ride sharing had severe impacts on the ride hailing and public transit markets, reducing the attractiveness of automated ride hailing. At the same time, the growing need for package delivery and concerns about person-to-person contacts in home delivery settings accelerated interest in automated local package delivery. These trends accelerated consideration of the workforce implications of automated driving, particularly among heavily unionized truck and bus drivers.
- The trends noted above also accelerated the growing recognition of how gradual the deployment rollout is likely to be for higher levels of automation. This motivated growing interest in identifying specialized use cases for automated driving that could be deployed earlier and generate nearer-term revenues, even if they represent modest-size markets such as ports, mines, or logistics hubs.

6.2 New Themes and Topics of Interest

- This year's meeting generated a higher level of interest and direct engagement from senior management at the U.S. DOT, including the welcoming remarks from the Secretary of Transportation and substantive participation throughout the meeting by senior DOT management officials.
- The leading participants from the industry appear to be converging on consistent frameworks for ensuring the safety of their automated driving systems. They are combining technical safety cases with consideration of organizational safety culture and attention to the full life cycles for their systems, for a robust and comprehensive approach to the safety challenges.

- The state and local government representatives appear to share common goals for their automated driving regulatory frameworks, even when those frameworks differ in specific implementations.
- The need for remote human support for higher levels of automated driving is receiving more attention, which is highlighting the safety, human factors and policy challenges of remote support to a greater extent than in the past.
- Infrastructure support for automated driving also attracted more attention than in the past, including both physical aspects (such as separations from other traffic) and digital support.

6.3 Public-Private Sector Interactions

- The international presentations and discussions highlighted significant differences in the public agency roles relative to the private industry roles. The private roles were most prominent in the Americas, the public most prominent in the Asia/Pacific, with Europe somewhere in the middle.
- The industry representatives appeared to be more willing than in the past to engage in a consultative process with other stakeholders to develop a broader consensus on regulatory approaches for automated driving.
- U.S. industry representatives expressed support for government playing the roles of developing measures of effectiveness for evaluating automated driving systems and for requiring "truth in labeling" on automation systems in order to enhance public confidence.
- U.S. industry representatives expressed reluctance to participate in public pilot projects that would require disclosure of significant technical or safety-related data, while recognizing that it will be necessary to share a certain amount of data to earn public trust. This tension regarding the appropriate amount of data sharing is likely to be an important issue for the future.

Part I: Public Sector and Policy Activities

Towards Social Deployment of Automated Driving Systems – SIP-adus Activities in Japan –

Yoichi Sugimoto[1(✉)] and Seigo Kuzumaki[2]

[1] Honda R&D Co., Ltd., 4630 Shimotakanezawa, Haga-Machi, Haga-Gun 321-3393, Tochigi, Japan
yoichi_sugimoto@jp.honda

[2] Toyota Motor Corporation, 1, Toyota-Cho, Toyota 471-8572, Aichi, Japan
seigo_kuzumaki@mail.toyota.co.jp

Abstract. This is a summary report on the latest SIP-adus activities. Cross-Ministerial Strategic Innovation Promotion Program (SIP) is led by the Japanese government. SIP-adus is one of the SIP themes, which is on connected and automated driving. The 2nd phase of SIP started in 2018, and SIP-adus is composed of the 4 pillars, which are technology development, public acceptance, international cooperation and field operational tests. As for technology development, it has the 4 focus themes, which are dynamic traffic environment information, traffic environmental data portal, virtual validation platform for safety assurance and evaluation methodology of intrusion detection system. In this report, dynamic traffic environment information and field operational tests are focused.

Keywords: automated driving · automated vehicles · connected vehicles · dynamic map · field operational test · ITS · V2I · V2N

1 Introduction of SIP-adus

1.1 SIP

Japan government promotes R&D policy under the concept of Society5.0 (Fig. 1). Through Society 5.0, it will be possible to achieve a society that can both promote economic development and find solutions to social issues by a high degree of convergence of cyber space and physical space. Cross-Ministerial Strategic Innovation Promotion Program (SIP) is one of propellants to realize Society 5.0. It began in 2014 as a five-year research program and the 2nd phase of SIP started in 2018 with 12 themes. SIP aims at promoting cross-sector government-industry-academia collaboration and intensive R&D from fundamental research to practical applications and commercialization.

1.2 SIP-adus

The project for automated driving systems for universal service (adus) was chosen by SIP as one of 11 research themes [1][2]. The governmental framework for promotion of

G. Meyer and S. Beiker (Eds.): ARTSymposium 2022, LNMOB, pp. 13–22, 2023.
https://doi.org/10.1007/978-3-031-34757-3_2

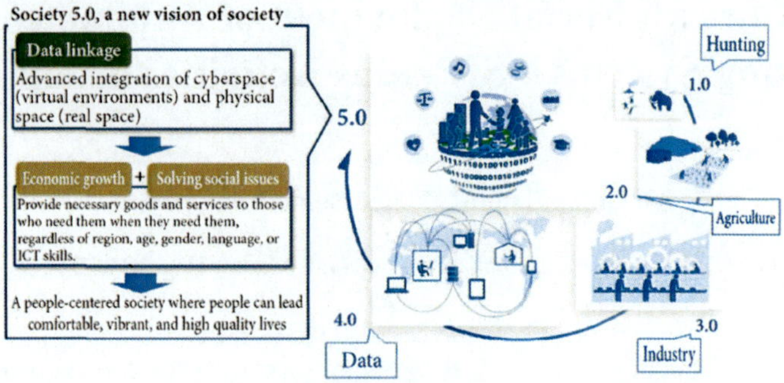

Fig. 1. Society 5.0

connected and automated driving systems includes four relevant ministries and agencies under the leadership of the Cabinet Secretariat and the Cabinet Office. Connected and automated driving systems, which provide benefits to our society, require collaborative efforts among government, industry and academia.

1.2.1 The 1st Phase of SIP-adus

When starting SIP-adus, reduction of traffic fatalities was set as the goal with the highest priority. Automated driving systems are thought to have large potential for reduction of traffic collisions. As it is challenging to cover all relevant technologies with limited resources available, the project prioritized five themes (Dynamic Map, HMI, cyber security, pedestrian collision reduction and next generation transport) as cooperative field technologies to be tackled.

1.2.2 The 2nd Phase of SIP-adus

The 2nd phase of SIP-adus is composed of the 4 pillars, which are technology development, public acceptance, international cooperation and field operational tests (FOTs).

Regarding technology development, the 2nd phase of SIP-adus has the 4 focus themes, which are traffic environment information, traffic environmental data portal, virtual validation platform for ADS (Automated Driving System) safety assurance and evaluation methodology of intrusion detection system (Fig. 2). Virtual validation platform for ADS safety assurance was reported in the previous volume of Road Vehicle Automation series [3]. In this report, traffic environment information is introduced in the following chapter.

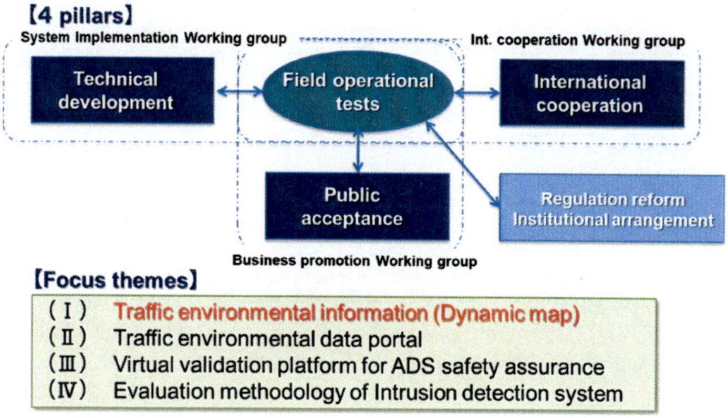

Fig. 2. Pillars and Focus Themes of SIP-adus Phase 2

2 Establishment and Utilization of Traffic Environment Information and Field Operational Tests

2.1 The Field Operational Tests in the Tokyo Waterfront Area

In order to realize advanced automated driving, it is necessary to build and utilize a framework for road traffic environment data, including a high definition 3D map. This database of road traffic environment data is called as a "Dynamic Map" and organized into a concept stratified by four layers: static data, semi-static data, semi-dynamic data and dynamic data. This Dynamic Map database is thought to be effective not only for automated driving vehicles but also for all other vehicles on the road.

In the 1st phase of SIP-adus, establishment of a static information platform was conducted. Based on the results of this R&D, distribution of the high definition 3D maps began in 2018, and it covered approximately 30,000 km of freeway nationwide in Japan. Subsequently, the world's 1st automated vehicle (Level 3) equipped with these maps was launched in 2021, and the maps are also being used by several automobile companies for their advanced safety driving support systems.

In the 2nd phase of SIP-adus, the "Road Traffic Environment Data Roadmap" (Fig. 3) was formulated to promote research and development for the establishment of a system for utilizing dynamic road traffic environment data, which is linked to high definition 3D map information. Standardization and practical use were advanced through the FOTs.

The FOTs were conducted from 2019 to 2022 in the Tokyo waterfront area (Fig. 4). The FOTs played important roles as places to openly confirm effectiveness and to identify issues through information distribution under real traffic environments on public roads. 29 entities participated in the FOTs, including car manufacturers, suppliers, universities and startup companies not only in Japan but also from overseas.

Since it was considered to provide traffic signal information, lane-level traffic information, merging lane assistance information, emergency vehicle information and detailed rainfall information with high priority, it was decided to generate and distribute such information in conjunction with high definition 3D maps for verification purposes.

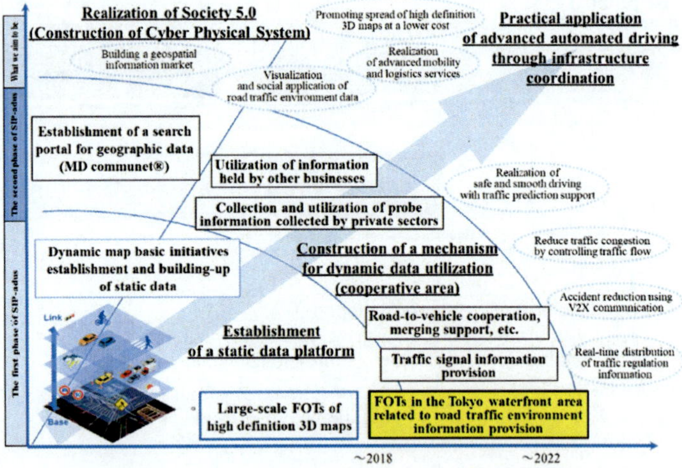

Fig. 3. Road Traffic Environment Data Roadmap

Fig. 4. Test Areas and Participating Entities of the FOTs

V2I (Vehicle to Infrastructure communication) and V2N (Vehicle to Network communication) were selected as communication methods, depending on data to be distributed (Fig. 5). The experiments showed that V2I is on track for practical application in traffic signal information distribution and that V2N has potential for road traffic environment data distribution [4].

2.2 Distribution of Traffic Signal Information

In the Tokyo waterfront area, V2I communication devices were installed at 33 signalized intersections, which distribute current signal information and remaining time to change colors. Also, the similar information was transmitted by V2N using cellular communication. Figure 6 shows traffic light recognition distances from an intersection by V2N,

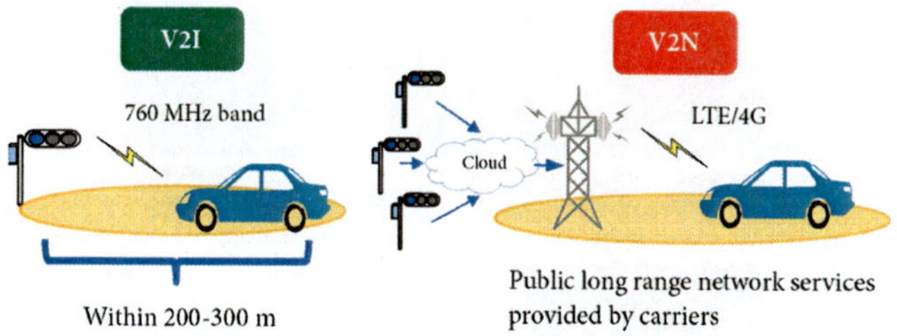

Fig. 5. V2I and V2N Communication System (for Traffic Signal Information)

V2I and an on-board camera. The result shows that the recognition distance using the on-board camera is approximately 150m and that the traffic signal information transmitted by V2I or V2N makes it possible to recognize signal status from farther distance, although all methods meet with the requirement for the recognition distance, which is assumed to be approximately 120m for vehicles to pass through an intersection. It was also found that recognition using an on-board camera becomes difficult in situations, where traffic signals are not physically visible or under severe environmental conditions such as shown in Fig. 7 and that V2I and V2N are considered effective in such situations.

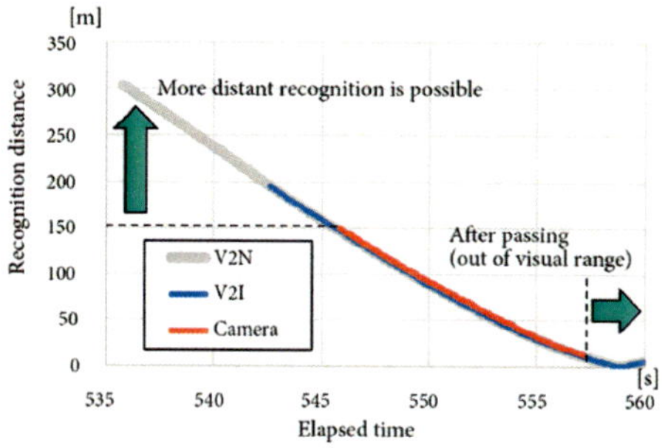

Fig. 6. Comparison of Signal Recognition Distances between V2N, V2I and an On-Board Camera [5]

In this project, a function to support smooth signalized intersection entry was also developed by utilizing information on remaining green time transmitted from V2I, and the effectiveness of this function was verified. In a so-called "dilemma zone", where a signal light color changes from green to yellow/red just before entering an intersection, sudden deceleration is unavoidable. It was confirmed that the function makes it possible to start deceleration before the actual signal light color changes, enabling smooth deceleration to stop at an intersection (Fig. 8). Since sudden deceleration in a dilemma zone can occur regardless of whether driving in an automated or manual mode as long as a decision to enter an intersection is based solely on current signal information, remaining green time information by wireless communication is considered to be useful [5].

Also, ISO/TS 19091 (Intelligent transport systems – Cooperative ITS – Using V2I and I2V communications for applications related to signalized intersections) was adopted

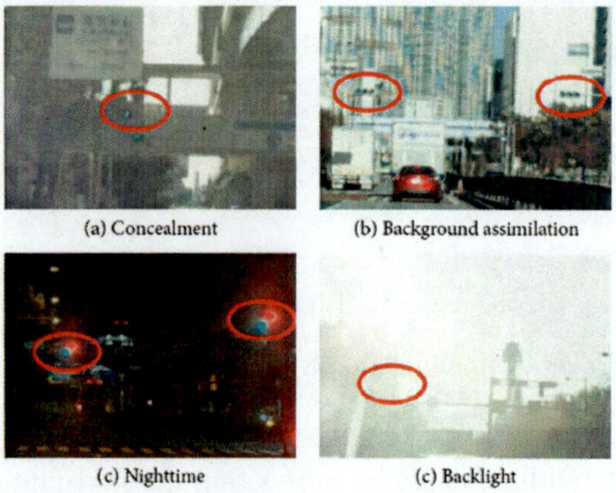

Fig. 7. Examples of Scenarios in Which Signal Recognition Becomes Difficult [5]

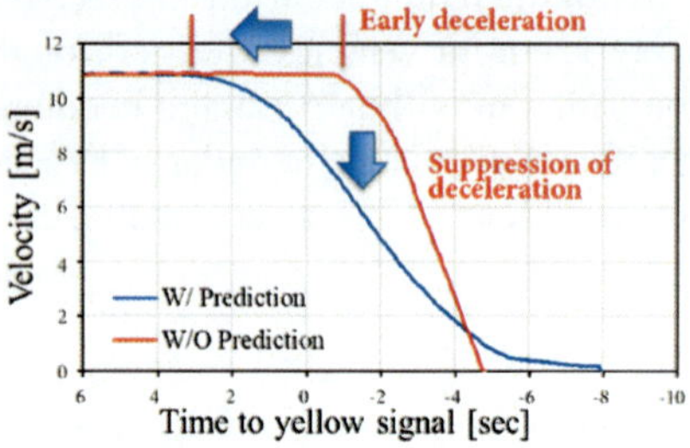

Fig. 8. Effectiveness of Deceleration Suppression in a Dilemma Zone using V2I Signal with Remaining Green Time Information [5]

as communication specification for the test vehicles. It was confirmed that it meets the requirements for automated driving systems.

2.3 Distribution of Traffic Environment Information

Since 2021, it was started to distribute traffic environment information by V2N, which can communicate with vehicles broadly. In addition to traffic signal information, lane-level traffic information, emergency vehicle information and detailed precise weather information were verified in the FOTs (Fig. 9). Those are expected to be effective for vehicles to prevent possible conflicts.

The overall flow of the provision of lane-level road traffic information is assumed to be the collection of data from probe operators, followed by the provision of the generated information to individual vehicles through the same probe operators (Fig. 10). Therefore, the component technologies that need to be considered for cooperative area of the lane-level road traffic information generation can be broadly classified into the following five areas: (1) data aggregation from probe operators, (2) integration of data from multiple information sources, (3) generation of lane-level road traffic information, (4) conversion to data capable of representing locations and (5) data distribution.

Figure 11 shows an overview of the traffic information generation in target use cases. Congestion tail information (A-1: branching support) is intended to be provided

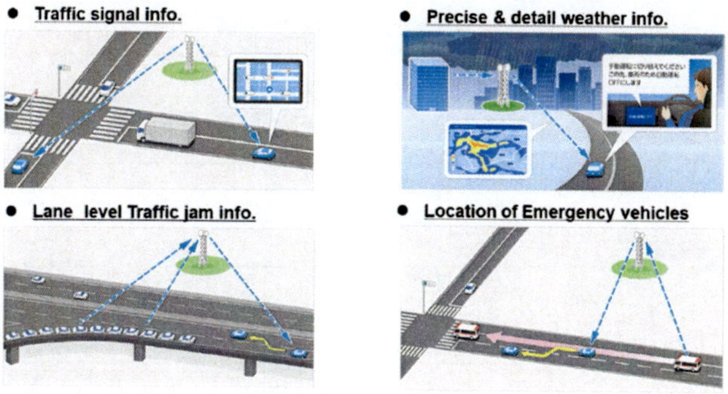

Fig. 9. Traffic Environment Information Distributed in the FOTs

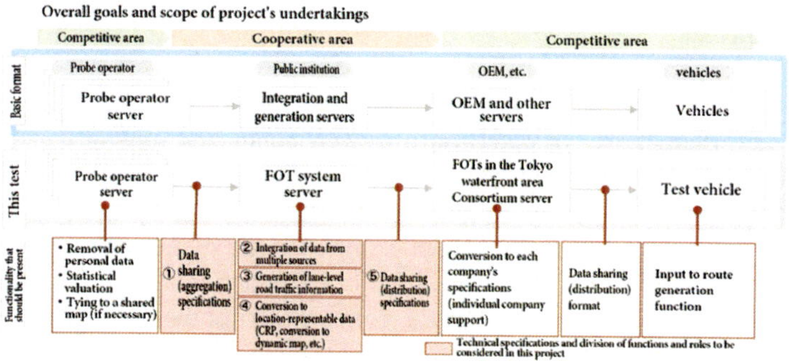

Fig. 10. Overall Flow and Scope of the Project on Lane-Level Traffic Information [6]

at branches. In the case of lane-by-lane congestions, the system estimates congested lanes based on speed information in each direction at a branch and provides information on congestion tails. Congestion tail information (A-2: passing support) is intended to be provided at merging points. The system estimates congested lanes based on turn signal information and provides information on congestion tails. Obstruction information such as traffic collisions, fallen objects etc. (B) is intended to be provided for all road segments. The information generation and the provision logic are the same as those of congestion tail information (A-2: passing support), but when there is little traffic and congestion does not occur, the system uses turn signal information to estimate lanes with an obstruction and provides information on locations of the obstruction.

The FOTs were conducted on two Metropolitan Expressway routes to investigate effectiveness of the information and to identify any further issues. As a result, it was able to obtain results regarding the amount of information that can be collected and the possibility of generating lane-level road traffic information using probe data, with a view toward practical application. Issues to be addressed for practical application include establishment of a sustainable operation system, data infrastructure development, maintenance of the system and clarification of a roadmap for service expansion [6].

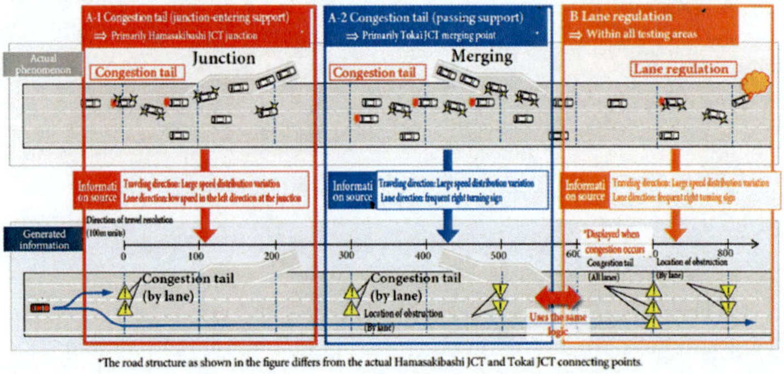

Fig. 11. Overview of Information Generation in the Target Use Cases [6]

3 Conclusions

In the second phase of SIP-adus, the R&D on the utilization of road traffic environment data and the FOTs in the Tokyo waterfront area were conducted, aiming at clarifying the requirements for road traffic environment data necessary for realization of automated driving and establishing a system for information generation and provision.

The technical prospect on practical application of traffic signal information provision through V2I was verified, which is necessary for introduction of automated driving on surface streets. The V2I environment in the Tokyo waterfront area will be passed on to and used for the next step of research.

In order to meet future needs for distribution of information from area-covering network, the R&D and the FOTs on various types of traffic environment information by V2N were also conducted. The effectiveness of the system and the potential for utilizing the information were demonstrated. The requirements and the issues for realizing the system were identified, and implementation of the system was proposed. It is expected that the results of efforts in the 2nd phase of SIP-adus will be utilized toward practical applications of road traffic environment data.

The reports by project and by fiscal year can be downloaded from the page linked from the "Research and Development" tab of the SIP-adus website. The English versions of the summary reports are available on its English site. Those who wish to refer to the reports can freely download the PDF files [7].

4 Next Steps

4.1 The 3rd Phase of SIP

The current 2nd phase of SIP will end in fiscal year 2022, and it was decided that the 3rd phase of SIP would start from fiscal year 2023. In order to fully consider the issues to be addressed in the 3rd phase of SIP in advance, 14 areas have been identified as candidate themes, including 10 areas corresponding to the 10 social visions listed in the Sixth National Strategy as “Society 5.0” that Japan is aiming at. Among those, there is a theme on “Smart Mobility Platform”.

The Task Force was formed to conduct feasibility study on possible sub-subjects of Smart Mobility Platform, and it was decided to study the four sub-subjects based on RFI (Request For Information) responses, which are (1) redefining mobility services and social implementation, (2) data platform supporting mobility services, (3) infrastructure strategy to support mobility services and (4) strategies for the social implementation of mobility services. The current plan for each sub-project is to identify core research themes and to conduct technical feasibility studies.

These studies are expected to be advanced efforts toward a smart mobility society that include organic and flexible linkage with various means of transportation and traffic environments as well as automated driving, while taking advantage of the results of SIP-adus [8].

4.2 RoAD to the L4

Another relevant project on automated driving in Japan is RoAD to the L4 (Project on Research, Development, Demonstration and Deployment of Automated Driving toward the Level 4 and its Enhanced Mobility Services), which started in 2021. It is a 5-year project led by METI (Ministry of Economy, Trade and Industry) and MLIT (Ministry of Land, Infrastructure, Transport and Tourism). The project has the 4 themes shown in Fig. 12. Theme 1 and 2 aim at realization of Level 4 mobility services at 50 locations or more by 2025. Theme 3 is to put Level 4 automated driving trucks into practical use in 2025 or later. Theme 4 is to realize infrastructure cooperation such as V2I and V2N communication to support automated driving in urban mixed traffic situations.

RoAD to the L4 aims to socially implement the results of theme 1 in society and to launch automated driving services in fiscal 2023 as a leading example of this project. For the other themes, the plan is to conduct necessary studies and preparations during the first two years, to start FOTs in the third year and then to realize social implementation after undergoing review for various criteria. For these plans, it is also expected to make maximum use of the results of SIP-adus, such as the traffic signal information distribution FOTs in Tokyo waterfront area, the study of communication methods that realize cooperative automated driving and the study of public acceptance, with the aim of early social implementation of level 4 and other automated driving mobility services [9].

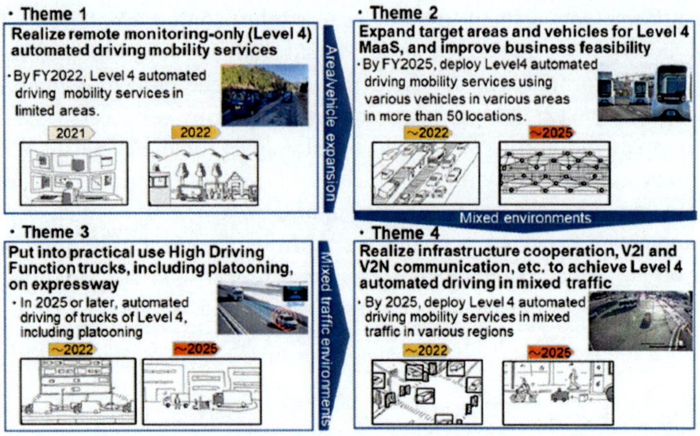

Fig. 12. Themes of RoAD to the L4

References

1. Amano, H., Uchimura, T.: A national project in Japan: Innovation of Automated Driving for Universal Services. In: Meyer, G., Beiker, S. (eds.) Road Vehicle Automation 3, Lecture Notes in Mobility, pp. 15–26. Springer, Cham (2016). https://doi.org/10.1007/978-3-319-40503-2_2
2. Amano, H., Uchimura, T.: Latest development in SIP-Adus and related activities in Japan. In: Meyer, G., Beiker, S. (eds.) Road Vehicle Automation 4. Lecture Notes in Mobility, pp. 15–24. Springer, Cham (2018). https://doi.org/10.1007/978-3-319-60934-8_2
3. Kuzumaki, S., Inoue, H.: Development of Driving Intelligence Validation Platform (DIVP®) for ADS safety assurance. In: Meyer, G., Beiker, S. (eds.) Road Vehicle Automation 9, pp. 13–22. Springer International Publishing, Cham (2023). https://doi.org/10.1007/978-3-031-11112-9_2
4. Hiyama, S., Minakata, M.: Establishment and Utilization of Traffic Environment Data and the Tokyo Waterfront Area Field Operational Tests (Overview), SIP 2nd Phase: Automated Driving for Universal Services Final Results Report (2018–2022) (2023). https://en.sip-adus.go.jp/rd/
5. Suganuma, N., et al.: Research on the Recognition Technology Required for Automated Driving Technology (Levels 3 and 4), SIP 2nd Phase: Automated Driving for Universal Services Final Results Report (2018–2022) (2023). https://en.sip-adus.go.jp/rd/
6. Ichikawa, H., Takenouchi, A., et al., Technological Development for Lane-level Road Traffic Information Using Probes Vehicle Data, SIP 2nd Phase: Automated Driving for Universal Services Final Results Report (2018–2022) (2023). https://en.sip-adus.go.jp/rd/
7. SIP 2nd Phase: Automated Driving for Universal Services Final Results Report (2018–2022) (2023). https://en.sip-adus.go.jp/rd/
8. Ueki, K.: SIP-adus Achievement as Heritage and Next Step Next Phase of SIP, SIP 2nd Phase: Automated Driving for Universal Services Final Results Report (2018–2022) (2023). https://en.sip-adus.go.jp/rd/
9. Yokoyama, T.: SIP-adus Achievement as Heritage and Next Step RoAD to the L4, SIP 2nd Phase: Automated Driving for Universal Services Final Results Report (2018–2022) (2023). https://en.sip-adus.go.jp/rd/

Regulatory Framework and Safety Demonstration Principles in France

Elsa Lanaud(✉) and Xavier Delache

French Ministry for Transportation, 1 Place Carpeaux, 92055 La Défense, France
{elsa.lanaud,xavier.delache}@developpement-durable.gouv.fr

Abstract. Automated road transport mobility will develop only if fundamental conditions are fulfilled: acceptance by users and citizens, economic sustainability, contribution to a more sustainable mobility and last but not least, demonstration of its safety. Fulfillment of these conditions needs to be addressed by policymakers through regulations, standards, guidance, assessments and stakeholders' involvement. France built its regulatory framework on this balance by assuming safety will be the main factor for other conditions for the development of automated road transport systems to be reached. This paper presents how France has set up its national framework for the development of automated road mobility, integrating various policy challenges in a safety-first based approach.

Keywords: French regulatory framework · national strategy · automated road transport system · safety demonstration · scenario-based approach

1 Introduction

Vehicles' automation and more precisely transport systems' automation is developing through increasing use cases, diversified functionalities and driving environments. In this context, one of the main challenges for public authorities is to set the right balance between innovation on one hand, and road safety and security concerns on the other. Policy actions for the development of automated mobility gather work forces on public and user acceptance, sustainability, economic sustainability, skills evolution, and last but not least safety demonstration. Technical regulation including technical guidance on safety demonstration and validation remain the key instrument to set the balance between innovation and road safety and security, both at national and international levels.

France has been an active stakeholder in the international scene since 2017, first by proposing to develop a technical regulation addressing the challenges of automated driving at the UNECE/WP29 level [1]. In the same time, France published its first road map for the development of automated road public transport [2]. This document already presented the will of France to equip itself with a strategy for the development of automated road mobility, supported by a collaborative ecosystem between the administration and the French industry.

French framework for the development of automated road mobility is based on a significant state-of-the-art to build a comprehensive and holistic approach. Following the

G. Meyer and S. Beiker (Eds.): ARTSymposium 2022, LNMOB, pp. 23–39, 2023.
https://doi.org/10.1007/978-3-031-34757-3_3

examples of its counterparts from the United States [3], Japan [4], Germany [5] or United Kingdom [6], France adopted its national strategy for automated road mobility [7]. This strategy aimed at building the regulatory framework for the development and then deployment of automated road mobility systems while reinforcing innovation through a consolidated testing regime.

The French approach is based on three principles – safety, progressivity, and acceptability. The French regulatory framework has put the system (beyond the vehicle itself), including remote capabilities and the service at the core of safety demonstration. By doing so, this approach has aimed to optimize the complementarities between vehicle type-approval, which is at the core of the European approach, and the need to fine-tune vehicle functionalities with predefined locations where those vehicles will provide a scenic (and vice-versa). The systemic approach considers the global system not only as a combination of each component but also as a whole, both independent and interacting with its environment. By its complexity and its vocation to provide interactions between the vehicle and human people – as well as passengers, other road users or a supervision center, an automated road transport system becomes critical for road safety. Safety as a pillar is probably the most important one as it is the binder between all activities in the development of automated road mobility. It is on the one hand the guideline of the regulatory framework to preserve road users and passenger vehicles' integrity, in compliance with the Highway Code. On the other hand, it plays a great role on acceptability and in particular, in building confidence in automated driving systems; the more people would trust the system, the more they would use it. Finally, by designing a performance- and progressive-based approach, safety is a guiding principle of type-approval and validation of systems.

1.1 Defining a Global Approach for the Development of Automated Road Mobility

For the last decade, vehicles' automation has been developing rapidly, though increasing levels of automation and diversified functionalities and driving environments. Building on the momentum and success of prior and current research, and working both on automation and connectivity – more generally on intelligent transport systems (ITS), a certain number of countries decided to launch their own national strategy. These strategies reflect the views from an entire ecosystem made from public authorities, industries and academics gathered around the development of automated road mobility. Although these strategies fall under national priorities and do not aim at reflecting common views, it enables the international community to establish a global vision on automated and connected mobility by proposing guidelines on the work to achieve. Experts from around the world are now able to share their strategic views, their experiences and their vision to move forward on both regulatory and technical framework.

At the United Nation (UN) level, a dedicated group has been set up under Working Party on Automated/Autonomous and Connected Vehicles (WP29/GRVA/FRAV) that aim is to set and define requirements for the type-approval of automated and connected vehicles. At the European (EU) level, the Commission has launched a working group on the type-approval of fully automated vehicles. The European Regulation 2022/1426 entered into force on August 2022 [8].

Europe has adopted a strategy for the development of automated and connected road mobility in May 2018 [9], supplemented by the sustainable and intelligent mobility strategy of December 2020 [10]. This framework notably includes a major component of support for innovation.

In line with other countries initiatives, France launched its own initiative to build its national strategy for the development of automated and connected mobility. The national framework gets inked in 2014 with an industrial roadmap for the development of autonomous vehicle. After Law 2015-992 [11] allowing experimental testing of automated vehicles, France initiated to work on a regulatory framework at UNECE level [1]. Several months later in 2017, French administration presented strategic directions to develop automated and connected vehicle through the prism of industrial benefits for the entire mobility system [2]. This document was then submitted for a national consultation. In November 2017, the French government decided to provide France with a strategy that, while integrating with international and European work, mobilizes national public and private stakeholders around the objective to strengthen France's position among the leading countries in the development of automated vehicles, and to contribute to the new mobility policy [7]. To achieve that goal, the Government appointed Anne-Marie Idrac, former Minister, as High representative in charge of the French strategy for the development of autonomous vehicles. The strategy formulated in 2018 has laid out three principles of action – safety, progressivity, and acceptability – that are still in effect. It has set up cross-functional working and leadership frameworks, in a public-private ecosystem. The strategy was articulated around two main axes: preparing the legislative and regulatory framework for the deployment of automation; supporting research and experimentation.

Part of these national strategies is safety a very important aspect, becoming more and more important as work on automated road mobility becomes clearer. In all most recent versions of national strategies, safety appears as one of the main topic, if not the most important aspect to work on. Technical and research works focusing more and more on safety demonstration confirm this statement.

1.2 Safety Demonstration as Main Pillar Supporting the Development of Automated Road Mobility

Automated driving systems (ADS) with high levels of automation require to set commonly and widely accepted and applied high level safety rules and efficient validation framework to ensure safety. Reflections on the development of autonomous driving safety validation have been very active in the past five years, based on the consensus that existing validation approaches have to be significantly modified [12]. Academia, industries, standard-setting organizations and regulators have produced a significant stock of ideas and proposals, as well as national governments around the world issued position papers.

At the UN level, a dedicated group is working on safety validation tools and methods (VMAD). The objective at the UN level is to propose an regulation architecture that considers a vehicle systemic approach, a diversity of tasks sharing between driver and system, and the diversity of use cases. Initiating that work at the UN level opens the global approach that national or regional authorities would have to follow. At the EU level, the

Commission decided to continue working on safety framework through the establishment of interpretation document and guidelines for example on ODDs, scenarios, safety target, safety validation tools, periodic roadworthiness.

In 2021, USDOT lays out the will to develop safety-focused frameworks and tools to assess the safe performance of ADS technologies, and to conduct the foundational research and demonstration activities needed to safely evaluate and integrate ADS, while working to improve the safety, efficiency, and accessibility of the transportation system [13]. This last version of its national strategy for the development of automated mobility outlines that entities involved in the development and testing of automation technology have an important role in not only safety assurance of ADS-equipped vehicles, but also in providing transparency about how safety is being achieved. In the same time, a 2018 report [14] proposes a preliminary research to develop a testing framework through existing test methods and tools to formulate an appropriate, comprehensive testing architecture around a scenario-based approach. This work puts forward the benefits of a scenario-based approach that fit flexibility within the test architecture.

The Law Commission of England and Wales and the Scottish Law Commission have completed a comprehensive 4-year joint review of the legal framework for automated vehicles, focusing on automated vehicles' safety and ascribing civil and criminal liability in the event of incidents [15]. U.K. is an active stakeholder in the field of safety demonstration, through its normative publications with the British Standards Institution (BSI). In that context, safety is part of all actions led by the Connected and Cooperative Automated Vehicle (CCAV) Code of Practice [16]: it aims at increasing public confidence by recommending that trialing organizations publish their safety case and educate the public on the potential benefits of a driverless transportation model. BSI has then developed and published a number of standards relating to connected and automated vehicles (CAV) with the aim of providing a set of industry standards and guidelines that stakeholders can use to take a safety-focused approach to CAV trialing through its BASI CAV Standards Programme. Part of all this normative content, it is possible to site among others PAS 1889 [17] that specifies requirements for a structured natural language format for test scenario definition of an automated driving system (ADS) Level 3 and higher.

In Germany, a great amount of work has been done on safety assurance for highly automated driving within the PEGASUS project, more particularly on driving scenarios [18]. Safety assurance has become a great deal because of the importance of ensuring an adequate safety level, compared to the current level with conventional cars driving by humans. The PEGASUS project initiated work on traffic scenarios to test vehicles' performance. There is a need to work on validation methods because current tools do not cover all activities and behaviors that will engage "robots" as we can consider automated vehicles as machines [18]. Considering scenarios means considering severity and frequency estimations in the validation process to put into service, systems that will not introduce more risks than conventional vehicles. In other words, the aim is to reduce the potential for accidents and to mitigate risks for passengers and other road users by operating like a defensive and attentive driver who consistently monitors the driving environment and responds appropriately and safely to changing conditions [19].

All that work highlights that sound, shared and transparent approaches for automated driving systems' safety validation are key to their development and acceptance. Validation will involve both public authorities and manufacturers. On the French side, the regulatory framework aimed at setting the foundation to the global safety demonstration and validation framework. Public administration, in close collaboration with the French Government, is establishing the entire safety methodologies to demonstrate systems' safety. This work is quite in line with the international framework, which would apply to the type-approval of vehicles equipped with an automated driving system. At the French level, the aim is to define methodologies and tools for safety demonstration and validation of automated road transport systems for the purpose of a transport service. These methodologies are built based on recommendations from the French industry [20]. Validation should then combine two main axes: a process-centered axis, to be mainly scrutinized by public authorities through audit of conception and validation methods; and, a performance-centered axis, to be mainly scrutinized by public authorities through tests. The efficient combination of these two-axis strongly relies on the management of driving scenarios for the conception and validation of automated driving systems.

2 Regulatory Framework in France Relies on the Safety Principle

The French regulatory framework on automated mobility entered into force in September 2022. With the amendment of the Vienna Convention on road traffic supports the first one. This framework is articulated with the EU regulation 2018/1426 [7]. This framework is two-fold: ordinance and decrees on one side, methodological and technical guidance on the other side.

This framework is embedded in a national strategy addressing other policy issues raised by the development of automated road mobility: acceptance and ethics, involvement and empowerment of local authorities, impact assessment, skills and training, support to research and innovation.

France has set up a national program on automated driving in 2015, led by the industry, under the New Industrial Action Plan ("Nouvelle France Industrielle"), with three main priorities: foster dialogue among different sectors of the industry contributing to automated driving development, identify common interest technology blocks, and identify regulatory gaps. This program has been organized in a matrix approach, i.e. by use cases (individual cars, public transport, freight and logistics) crossed with technological and regulatory issues.

2.1 National Regulatory Framework for the Development of Automated Road Mobility

2.1.1 National Strategy

The French strategy for the development of automated and connected road mobility has been launched in 2018 [6] and revised first in 2020 [21]. The strategy was built from contributions from the entire French ecosystem as the research community through the "Nouvelle France Industrielle" consortium [22]. Among of these contributions, a certain number of French technical and research organizations shared their views on priorities to develop the automated mobility as INRIA [23], IFSTTAR [24] and VEDECOM [25]. This strategy has been updated more recently in the end of January 2023 [26].

This update landmarks a new stage: it explicitly takes into account connectivity issues, and enlarges the scope to mobility services made possible by automation and connectivity. Under the aegis of the Ministers of Ecological Transition and Territories, Economy, Interior and Overseas Territories, Transport, Industry and Digital, this strategy is being developed in close cooperation with a very thriving cross-sector private ecosystem. It is articulated with European and international work. All of the work involved in defining and implementing this ambition is entrusted to a senior official, former minister Anne-Marie Idrac.

Various collective work tools were mobilized in the second half of 2022 to build the strategy update with all stakeholders, focusing on four priority actions:

- Prioritize and coordinate connectivity systems and data exchange deployments. This involves defining common priorities for connectivity use cases among all stakeholders, with regard to road safety and operations issues, and economic benefits. This action will have to pay particular attention to the needs of connectivity and on-board intelligence for automated public or shared transport, in order to accelerate their deployment.
- Finance investment projects in industrial supply of automated road mobility, ambitious service pilots, or first commercial deployments, in particular via France 2030 and by mobilizing European credits. The objective is to extend the measures put in place as part of France 2030, in order to finance the development and industrialization of automated and connected vehicles and their components, as well as the first commercial deployments of passenger transportation services based on these vehicles.
- Supporting volunteer local authorities and operators in the deployment of passenger services. The objective is to make the regulatory framework a deployment facilitator for local authorities, transport operators and site managers. The preferred tool will be a resource center to share experiences and guide the design and evaluation of deployment projects and the application of the safety demonstration framework. The 2030 target is for 100 to 500 services without an onboard operator, i.e. several thousand vehicles.
- Finalize the legal framework for automated freight and logistics. The objective is to enable the development of use cases by creating the necessary framework for traffic on open roads, in addition to the existing framework that already allows operations on closed sites.

Actions cover a relatively short period of time (2023–2025), in order to be able to act quickly and reevaluate needs according the context's evolution, notably use cases' technical and economic feasibility.

2.1.2 Regulatory Framework

The regulatory framework for the deployment of automated road mobility comes from the "Loi d'orientation des mobilités" (LOM) of December 24, 2019, which defines roles and responsibilities of automated systems and drivers or human supervisors, as well as the processes for safety demonstration [27]. It allows the deployment of systems without driver on board, subject to safety demonstration on a predefined route or traffic area, as well as remote supervision and intervention functions.

The regulatory package for the deployment of automated road transport systems is composed of Ordinance no. 2021-443 from April 14, 2021, supplemented by Decree no. 2021-873 from June 29, 2021. The French ordinance defines the roles and responsibilities of automated systems and drivers or human supervisors, as well as principles concerning automated road transport systems. These principles include the commissioning decision by the service organizer, existing conditions of use set by the manufacturer, responsibilities for both the service organizer and the operator in the case of accident, and supervisors' responsibilities. The architecture defined by Ordinance no. 2021-443 is presented in Table 1.

Table 1. Architecture of Ordinance no. 2021-443 from April 14, 2021

Use case	Case A: Driver	Case B: Supervisor
Case I: Automated system that has to ask for a takeover for certain types of hazards during a maneuver, in its operational design domain	• Type-approved vehicle • Driver shall be able to takeover control when requested by the system • = partially automated vehicle	Not allowed
Case II: Automated system designed to answer to every hazards without requesting a takeover in its operational design domain	• Type-approved vehicle • Driver shall be able to takeover control in the case of ODD ending • = highly automated vehicle	• Type-approved vehicle • Only in a transport system (passenger, freight) • In a predefined route/zone • After decision from the service organizer based on safety demonstration • Supervisor able to intervene depending on the system's conditions of use • = fully automated vehicle

The Decree no. 2021-873 from June 29, 2021 defines the concept of "automated road system". More particularly, it defines the following notions:

- vehicle with delegated functions (partially, highly and fully automated) = vehicle equipped with an automated driving system
- "Partially" automated vehicle = must do a take-over request to respond to some traffic hazards or failures during a maneuver
- "Highly" automated vehicle = can respond to any traffic hazard or failure (within its operational design domain), without doing a take-over request during a maneuver
- "Fully" automated vehicle = can respond to any traffic hazard or failure, without doing a take-over request during a maneuver, and used in automated road transport systems with remote intervention capability.

- operational design domain = the operating conditions, including (but not limited to) environmental, geographic and time specifications, under which an automated driving system is designed to exert dynamic control of the vehicle and to inform the driver
- minimum risk maneuver = stopping the vehicle in a situation of minimal risk to its occupants and other road users, and automatically performed by the automated driving system, following a hazard not foreseen in its operating conditions, a serious failure or, in the case of remote intervention, a failure to acknowledge a maneuver requested by the system
- emergency maneuver = maneuver automatically performed by the automated driving system in the event of an imminent risk of collision, with the aim of mitigating or avoiding it
- It also defines the specific notion of "automated road transport system".
- automated road transport system = a set of highly or fully automated vehicles, and technical devices allowing remote intervention or safety, deployed on predefined routes or zones, and complemented by operating and maintenance rules, for the purpose of providing a passenger road transport service
- remote intervention = only within an automated road transport system, can activate, deactivate the system; give instruction to the system to perform, modify, interrupt a maneuver; acknowledge maneuvers proposed by the automated driving system; and choose, modify the planning of a route or stop points

Any automated road transport system is subject to conditions of use, which specify:

- operational design domain of vehicles used in the system
- technical design domain of the technical system
- conditions under which a minimum risk maneuver is activated by the automated driving system
- conditions under which an emergency maneuver is activated by the automated driving system
- conditions under which an authorized person may, outside the vehicle, give the instruction to carry out, modify, interrupt or acknowledge a maneuver
- description of maneuvers that can be performed outside the vehicle
- description of acknowledgment conditions for maneuvers proposed by the system and which can be acknowledged remotely

Furthermore the French decree defines safety demonstration principles, following the "globally at least equivalent" (GAME) principle: "*Any automated road transport system or any part of an existing transport system shall be designed, put into service and, where appropriate, modified in such a way that* ***the overall safety level with regard to users, operating staff and third parties is at least equivalent to the existing safety level or to the safety level resulting from the implementation of the systems or subsystems providing comparable services or functions****, taking into account the state of the art, the feedback from experience concerning them and the reasonably foreseeable traffic conditions on the route or in the traffic area concerned*".

The vehicle, integrated in a transport system, has to be type-approved either at the EU under the EU 2022/1426 regulation or at the French level as a specific vehicle category (French shuttle category). To be deployed in a transport service as an automated road

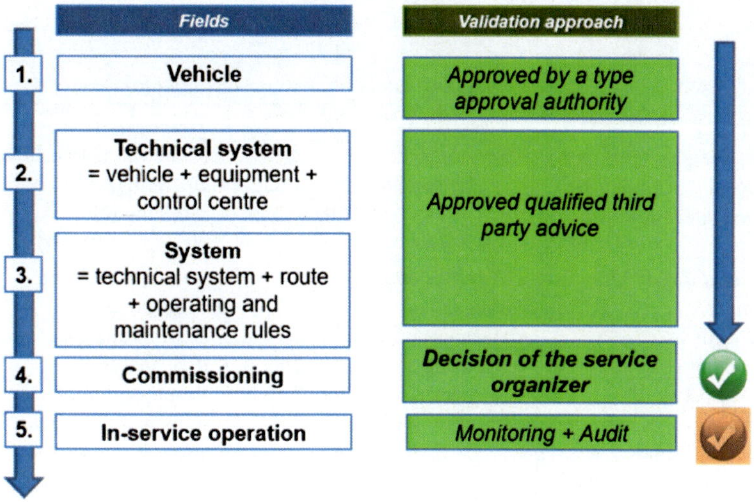

Fig. 1. French safety demonstration architecture for automated road transport systems

transport system, it has to be validated at the national level on a predefined route or zone. Figure 1 shows the global architecture of French regulatory framework and presents the links established between successive validation processes and approval bodies.

The safety demonstration is established before the automated road transport system is put into service; more precisely the aim is to validate that the system is able to respond to any traffic hazard in its operational design domain and is not going to add new safety risks.

The safety demonstration is composed of three different files, supplemented by third qualified bodies advices. The Table 2 below shows the content of each file.

The French decree has been complemented in 2022 by three orders to precise the role and qualification of third parties, and remote operators authorization.

2.2 Safety Demonstration

In 2018–2019, in parallel with the preparation of the French national strategy, the French industry program has been restructured, putting forward safety validation as the core task, renamed "France Vehicles' Autonomes" (FVA). Back-to-back groups have been set up between French administration and FVA on safety validation, responsibility, driving code, vehicles identification, and testing.

At the national level, various guidelines have been published on safety demonstration methods, detailed below. The French safety demonstration approach is based on a set of bricks, following a performance-based approach.

Table 2. Validation domains and safety files

Technical system	Preliminary safety file	Safety file
1. ***Description*** • Vehicle category/type and operational design domain • Functional decomposition	1. ***Description*** • Service/route characterization • Operating, maintenance and servicing rules	1. ***Description*** • Route characterization • Operating, maintenance and servicing rules
2. ***Analysis and risk management*** • Failure analysis • Allocation of safety objectives between sub-systems • Safety demonstration, tests and simulations performed	2. ***Analysis and risk management*** • Hazards analysis • Risk analysis • Responses and safety framework (GAME)	2. ***Analysis and risk management*** • Hazards analysis • Risk analysis • Responses and safety framework (GAME) • Tests done on the route
3. ***Specifications for implementation*** • Generic route description • Generic need for infrastructure (connectivity) • Tests needed for route commissioning	3. ***Specifications for implementation*** • Equipment (including connectivity) • Traffic police rules	3. ***Specifications for implementation*** • Response and safety plan

2.2.1 Main Principles of the Scenario-Based Approach

Safety demonstration activities in France are articulated around the scenario-based approach [28]. More precisely, the scenario-based approach is articulated with safety demonstration activities, in particular those resulting from safety operation (ISO 26262) or safety of the intended functions (ISO 21448), or from GAME approach.

As mentioned in the EU 2022/1426 [7], scenarios are the core of safety demonstration, acting as the link between all activities before putting automated road transport system in the market. Scenarios are also the core of safety assessment principles, built around four pillars:

- overall safety level, to evaluate the residual risk
- risk analysis completeness, to cover "reasonably foreseeable" critical situations, including failures and traffic hazards within the operational design domain
- specific mandatory scenarios, for which the system has to demonstrate its capabilities to respond safely without adding new risks for passengers and other road users (demonstrating it preserves the users' integrity)
- specific families of scenarios, for example detect and respond to certain types of other road users, emergency vehicles, environmental conditions

A major challenge of the safety demonstration approach is to ensure the greatest possible completeness in the risk analysis. The objective is, as the SOTIF standard reminds us, to limit the cases of unknown and serious risks (also known as "unknown unsafe").

This quest for scenario completeness is central. The requirement of "reasonably foreseeable" applies to it, to which the deductive and inductive approaches presented above contribute. These approaches must combine the search for system malfunctions and external traffic hazards. This search for completeness of events (malfunctions to traffic hazards) is fed by the combination of deductive approaches on possible causes (hazard to possible causes) or inductive approaches of failure modes (failure to hazards). The search for completeness is also fed by the search for route specific hazards.

The scenario approach is based on a method of "expert" generation, by combining scenario description axes (driving environments * nominal maneuvers * characteristics of collision precursor events), which leaves aside any objective of probabilization and does not rely on data from observation.

The main contribution expected from the scenario-based approach is thus to avoid the omission of certain types of scenarios in the quantified (probabilized) approaches of traditional risk analyses. Scenario generation [29] (in particular by combining axes) is therefore a first step, normally completed by a quantification step (frequency/severity). The robustness of the overall approach (scenario generation/quantified risk analysis) should therefore be based on the principle that scenarios resulting from the scenario generation stage are systematically included in the quantified analyses, even if it means qualifying them as "implausible" or qualifying their consequences as "not serious". The main objective of scenario generation is to reduce the range of "unknown" scenarios, not to assess their criticity.

In a complementary way to scenario generation and feeding, potentially integrated in quantitative methodology, the use of these scenarios has to be defined and framed [30]. Based on the VMAD NATM master document [31], France produced a methodological document on validation methods based on the scenario catalogue. The NATM pillars are closely linked to the scenario-based approach as the foundation for test relevance, to ensure a holistic and dense coverage of traffic scenarios.

2.2.2 Conception-Validation Principle for an Automated Road Transport System

All these activities are part of a global model to validate automated systems: conception activities, audit and performance validation [32]. The safety demonstration methodology, which can be structured in several phases, is an iterative process. In the automotive culture, the system development process is based on the so-called "V-cycle" model. In this model, system design is based on the identification of interacting subsystems. The V-cycle model covers all the development phases of a system component, from specification to validation. According to the literature, ISO 26262 and the SOTIF standard, the V-cycle can be structured in several phases from the system specification to safety demonstration at the system level as shown in Fig. 2. Safety demonstration is composed of system specification (1); system risk analysis (2); function risk analysis (3); system validation (4); and system's safety demonstration/file (5).

These activities, part of the V-cycle are combined in an iterative way, by adjusting the specifications through failure control measures (inductive approach) or overall performance (deductive approach). The new risk analyses applied to the adjusted

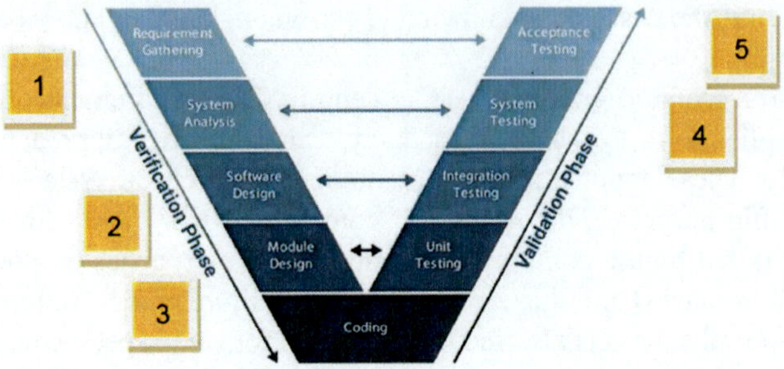

Fig. 2. V-cycle and main safety demonstration activities

specifications, eventually lead to a revision of the overall system specifications, until they converge towards an acceptable residual risk level for the system, in reference to comparable systems or services. The specifications, the risk level and the risk analysis documentation then constitute the substance of the safety demonstration file.

Some of the activities in the V-approach described above consume and/or generate scenarios, which are then specific to a given system. The particularity of the V-cycle model is that it is applicable to a system but also to systems of systems. Therefore, it can be used to validate safety at the vehicle level but also at the system level when integrating one or more vehicles in a technical system, as defined by the French decree. The integration of this technical system in an automated road transport service on a predefined route – considered as a system – is validated in the same way. In a schematic way, the validation of the safety of an automated road transport system, from the vehicle, system and service components deployed on a route, is an iterative process from an iterative model. The studied system is then respectively:

- the vehicle equipped with an automated driving system (delegated driving vehicle),
- the equipment external to the vehicle (technical installations deployed on the route, supervision),
- the technical system (vehicle with driving delegation and validated technical installations),
- the automated road transport system (technical system deployed on a predefined route with operating, maintenance and servicing rules).

2.2.3 Scenario-Based Approach Articulation with Other Safety Demonstration Activities

On the one hand, the design-validation approach produces and consumes scenarios: it essentially produces scenarios in the risk analysis stages; it uses scenarios in the various test and validation phases (at the component and system levels). This process aims to ensure that the scenarios are representative of the system in question, and in particular of its operational design domain. This articulation can be described as “endogenous”.

The link between scenarios and ODD is an important safety topic as ODD description is the starting point to build scenarios [33]. ODD is also more than just describing the system operational design domain as it could refer to the vehicle’s domain of use, to the technical system’s domain of use (as described in the French decree), or to the transport system's operational design domain, when deployed in a predefined route (as described in the French decree).

Route description for the deployment of an automated road transport system is thus closely linked to operational design domain description as to the scenario-based approach [34]. Furthermore, the search for ODD specific scenarios must be based on the search for scenarios specifically identified in the route safety analysis on which an automated road transport system is to be deployed. These scenarios are likely to either better filter the scenarios from rather "exogenous" sources, in order to eliminate those that are reasonably non-conceivable on the route, or, on the contrary, to see the appearance of risk factors specific to the route (e.g. particular geometries, visibility masks, points of accumulation of collision-generating events).

On the other hand, the scenario-based approach has the advantage of being able to provide a form of "exogeneity", useful in the safety demonstration, by using scenarios that can be derived from references external to the system. These scenarios must of course be filtered to ensure that they do indeed reflect "reasonably foreseeable" scenarios in the specific operational design domain, but their great advantage is that they contribute to limit the risk of omitting scenarios, that an approach that is too "endogenous" to the system could run.

The sources of scenarios that can be mobilized are diverse: those resulting from the design of the system and the risk analysis applied to the various components of the system; those resulting from observed data (driving, near misses, and accidents); those resulting from regulations, standards or codes of practice. The sources can be enriched by each other with the objective of completeness. Collaborative work, in particular the sharing of scenario databases, also contributes to enriching the sources.

When dealing with scenarios resulting from regulations, standards or code of practice, an important work has to be done to create and consolidate regulatory lists of scenarios that the system will have to "pass" to demonstrate its capabilities, more particularly to assess that the system will not create new safety risks. The first requirement the system shall be in line with is the respect of traffic rules. A large codification process has begun within the French safety demonstration framework, to be able to "translate", in an understandable wording, some requirements of the traffic rules. For example, the traffic code includes notions of good practice that are not easily understandable and reproducible by a machine. Moreover, the requirement under which the system shall respect the traffic rules applies to interaction scenarios with priority vehicles (such as emergency vehicles) and law enforcement officers [35].

Thus, the scenario-based approach when being integrated in a more quantitative approach, poses the question of safety assessment methods and tools to develop and justify an overall safety target. The definition of a safety objective for automated driving systems is one of the major issues in the regulation of these systems. This objective (generally expressed as a rate of accidents that can lead to death or injury per metric of use – km driven or hours of use), is intended in particular to allow the allocation of global safety requirements, defined at the system level, to the various subsystems or sub-functions that make it up. This approach of decomposition into sub-functions or sub-systems is indeed at the heart of the safety design and validation methods used by the economic actors (cf. the so-called "V" approach of the standards, ISO 26262, ISO 21448

SOTIF; or the GAME approach – Globally at least equivalent) or standard EN 50126 in the railway domain. In France, a preliminary work has be done on the national database to quantify the safety level on national roads [36]. This definition raises various types of questions from the point of view of public action, mainly concerning the appropriateness of defining guardianship values, their legal status and the authority that defines them, the scale (European or national or local) at which these objectives are set, and the way in which the road accident rate of conventional vehicles is used. For this study, only three use cases have been developed: individual car on motorways, individual car on national road categories, and public transport in an urban environment.

Finally, the design-validation approach cannot be detached from its operation and monitoring part, which generates feedback and thus enriches the sources of observation.

The Fig. 3 below presents the articulation of the scenario-based approach.

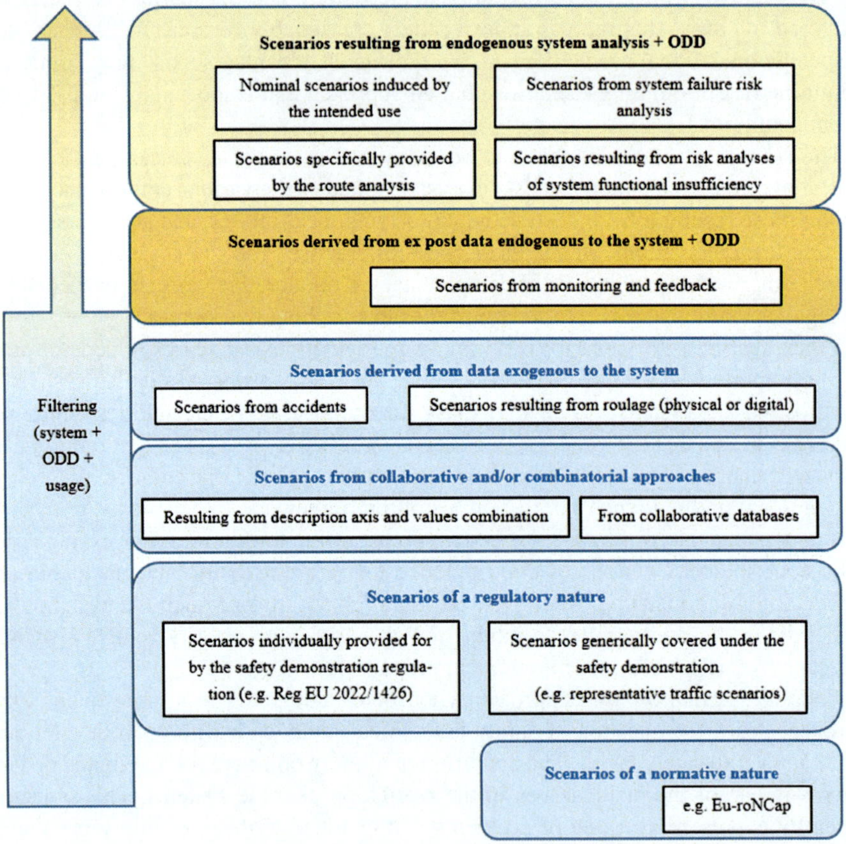

Fig. 3. Scenario-based approach articulation

3 Conclusions

As the French regulatory framework is now ready to enable the deployment of automated road transport system for passenger transport since 1 September 2022, the focus of forthcoming activities will be to provide guidance on safety demonstration methods and tools.

The safety demonstration framework is articulated around the scenario-based approach that enables the generation of traffic scenarios within the reasonably foreseeable conditions, which is the ink point of the overall safety demonstration framework. Work achieved on guidance already covers work on:

- ODD description and characterization [33],
- interaction scenarios with emergency vehicles and law enforcement vehicles [35],
- overall safety level and assessment tools [36],
- "globally at least equivalent" principle [37, 38],
- missions of approved qualified bodies [39],
- cybersecurity principles [40],
- route characterization [34].

The next step will cover:

- remote operators scenarios and functionalities,
- traffic rules scenarios,
- regulatory scenarios,
- use of feedbacks from accidents and incidents.

The aim of this continuing work is to switch from the generic and global scenario-based approach to more specific and use-case tailor-made guidance.

Work on guidance will highly involve local authorities, beyond industry eco-system. Some aspects will deserve larger consultations in conjunction with representatives of civil society, including representatives of users inter alia.

The French strategy is still based on these three founding pillars: safety, progressiveness and acceptability. The scenario-based approach is part of this logical process to develop automated road mobility in a safe manner, fed by a performance-based approach, taking into account societal aspects and user acceptance of these systems.

References

1. French Ministry for Transportation. International horizontal regulation of automated vehicles: preliminary framework considerations (working document), 60 (2017)
2. French Ministry for Transportation. Développement du véhicule automatisé: Orientations stratégiques pour l'action publique, 56 (2017)
3. Joint Program Office, U.S. Department of Transportation, ITS Strategic Plan 2015–2019, 83 (2014)
4. Japanese Ministry of Economy, Trade and Industry (METI), and Ministry of Land, Infrastructure, Transport and Tourism (MLIT), Panel on Business Strategies in Automated Driving "Action Plan for Realizing Automated Driving" Version 2.0, 58 (2018)
5. German Federal Government, Strategy for Automated and Connected Driving, 32 (2015)

6. Centre for Connected and Autonomous Vehicles, U.K. Government, Pathway to Driverless Cars: Consultation on Proposals to Support Advanced Driver Assistance Systems and Automated Vehicles – Government Response, 19 (2017)
7. French Ministry for Transportation, Development of Autonomous Vehicles: Strategic Orientations for Public Actions, 96 (2018)
8. https://eur-lex.europa.eu/legal-content/EN/TXT/PDF/?uri=CELEX:32022R1426&from=EN
9. https://eur-lex.europa.eu/legal-content/EN/TXT/PDF/?uri=CELEX%3A52018DC0283&from=EN
10. https://eur-lex.europa.eu/resource.html?uri=cellar:5e601657-3b06-11eb-b27b-01aa75ed71a1.0001.02/DOC_1&format=PDF
11. https://www.legifrance.gouv.fr/loda/id/JORFTEXT000031044385
12. Wagner, M., Koopman, P.: A Philosophy for Developing Trust in Self-Driving Cars, Road Vehicle Automation 2, Lecture Notes in Mobility, pp. 163–170. Springer (2015)
13. United States Department for Transportation, Automated Vehicles: Comprehensive Plan, 38 (2021)
14. National Highway Traffic Safety Administration, U.S. Department of Transportation, A Framework for Automated Driving System Testable Cases and Scenarios, DOT HS 812 623 (2018)
15. Law Commission of England and Wales, and Scottish Law Commission, Automated Vehicles: Joint Report, 315 (2022)
16. Pegler, L.: Connected and Automated Vehicles: A Review of the UK's Legislation and Good Practice. British Standard Institute White Paper (2022)
17. Centre for Connected and Autonomous Vehicles. Natural Language Description Abstract Scenarios for Automated Driving Systems – Specification, PAS 1889:2022. British Standard Institute (2022)
18. Winner, H., Wachenfeld, W., Junietz, P.: Safety Assurance for Highly Automated Driving – The PEGASUS Approach, Automated Road Symposium Sans Francisco (2016)
19. Mercedes-Benz, B.: Reinventing Safety: A Joint Approach to Automated Driving Systems (2018)
20. French Automotive Platform. Automated Driving Safety Validation: Proposals from the French Eco-system (Working Concept Paper), 30 (2020)
21. French Ministry for Transportation. The French Strategy for the Development of Automated Road Mobility 2020–2022, 12 (2020)
22. Nouvelle France Industrielle. Objectifs de recherche: Nouvelle France Industrielle Véhicule Autonome, 14 (2015)
23. National Institute for Research and Artificial Intelligence. Véhicules autonomes et connectés: Les défis actuels et les voies de recherche, 49 (2017)
24. French Institute for Science and Technology in Transport, Urban Planning and Networks. Le véhicule autonome: Enjeux et priorités pour la recherche, 27 (2016)
25. VEDECOM Institute. "Véhicule Autonome: Accompagner la transition – Perspectives d'usages et enjeux pour les différents acteurs", 14 (2015)
26. French Ministry for Transportation. Development of automated and connected road mobility – state of play, challenges and actions for the national strategy, 8 (2023)
27. https://www.ecologie.gouv.fr/loi-dorientation-des-mobilites
28. French Ministry for Transportation. Démonstration de sécurité des systems de transports routiers automatisés: articulation des différentes activités autour de l'approche "scenarios" – Document méthodologique, 27 (2022)
29. French Ministry for Transportation. Safety demonstration of automated road transport systems: contribution of driving scenarios: generation, feeding and enrichment of scenarios (2022)

30. French Ministry for Transportation. Utilisation des scenarios pour la demonstration de la sécurité des systems de transports routiers automatisés – document méthodologique, 20 (2023)
31. VMAD, New Assessment/Test Method for Automated Driving (NATM) – Guidelines for Validating Automated Driving System (ADS) (2021)
32. French Ministry for Transportation. New validation approaches for automated driving safety – French Views, UNECE/GRVA/VMAD (2019)
33. French Ministry for Transportation. Vehicles and automated transport systems: first principles and questions for the definition of the ODD (methodological framework document – initial version) (2022)
34. Tattegrain, H., Hedhli, A., Dusserre, A., Jouve, P.: STPA: Analyse de sécurité des parcours prédéfinis (2022)
35. French Ministry for Transportation. Safety demonstration of automated road transport systems: method for developing interaction scenarios with priority vehicles benefiting from ease of passage and interactions with law enforcement officers (methodological report) (2022)
36. French Ministry for Transportation. Safety validation of automated road transport systems: clarification through the analysis of accident data (study report) (2022)
37. French Ministry for Transportation. Automated road transport system – "GAME" principle globally at least equivalent (implementation guide) (2021)
38. French Ministry for Transportation. Automated road transport system – technical guide related to "GAME" demonstration (technical guide) (2022)
39. French Ministry for Transportation. Systèmes de transport routier automatisés – Mission de l'organisme qualifié agréé pour l'évaluation de la sécurité et pour l'audit de sécurité en exploitation des STRA (Guide d'application) (2022)
40. French Ministry for Transportation. Systèmes de transport routier automatisés – Guide d'application relative à la cybersécurité pour les STRA (Guide d'application) (2022)

Public Sector Integration of Connected and Automated Vehicles: Considerations, Benefits and Sharing Data Across Borders

Lisa Miller(✉)

The Eastern Transportation Coalition, 5000 College Park Avenue, College Park, MD 20740, USA

limiller@tetcoalition.org

Abstract. As the transportation landscape changes, accommodations for Connected and/or Automated Vehicles (CAV, AV and CV) must be considered. For this paper, CAV is referring to the technology available on the roadway now and what is anticipated soon – mostly the SAE International Levels 1 and 2 (SAE Levels of Driving Automation™ Refined for Clarity and International Audience). CAV considerations for freight and multimodal will not be discussed in this paper. The dynamic nature of CAV Levels 3, 4 and 5 leaves much to research. Public sector agencies in the United States are shifting focus to automated mobility, but with an increasing focus on near-term market ready technologies. Public outreach and education, coordination with Original Equipment Manufacturers (OEMs), policy and legislation and cross-border collaboration are also integral considerations.

Keywords: connected vehicles · automated vehicles · V2X · V2I · V2V · vehicles · safety · coordination · collaboration · membership · coalition · transportation · trends · emerging · technology · technologies · dynamic short-range communication · latency · advanced driver assistance systems · mobility · operations · connectivity · original equipment manufacturers · public sector · private sector

1 Introduction

The breakout session was intended to bring together industry experts for a state of practice discussion. The session moderator selected the panelists to provide a broad cross section covering the automotive manufacturing perspective, a corridor coalition perspective, a European perspective, and a data provider perspective.

Session attendees were able to hear the state of the practice for CAV and ask questions of the plenary speakers. The objective was to further learn in the industry and to collaborate on testing and deployment of CAV technologies to improve roadway safety and mobility.

The possibility of safer roads with better travel time reliability and mobility is more elusive than it should be, so what is missing? Engineers, planners, and other transportation professionals have been chasing an end to roadway fatalities for decades. Transportation professionals are familiar with the ever-present annual statistic of approximately

G. Meyer and S. Beiker (Eds.): ARTSymposium 2022, LNMOB, pp. 40–50, 2023.
https://doi.org/10.1007/978-3-031-34757-3_4

42,915 fatalities on roadways in the United States (NHTSA, 2022) during 2021. In 2020, the Insurance Institute for Highway Safety (IIHS) data shows 38,824 fatalities (2022). The United States Department of Transportation (USDOT) and the National Highway Traffic Safety Administration (NHTSA) provide comprehensive statistics on these fatalities (Table 1, 2022). These figures illustrate that despite 20 years of transportation safety and zero fatalities being the target of transportation organizations nationwide, the progress to reduce fatal crashes is clearly insufficient (Fig. 1).

Table 1. National Highway Traffic Safety Administration (NHTSA) 2022 Fatality Statistics.

Year	Deaths	Crashes	Motor vehicles
2000	41,945	37,526	57,594
2001	42,196	37,862	57,918
2002	43,005	38,491	58,426
2003	42,884	38,477	58,877
2004	42,836	38,444	58,729
2005	43,510	39,252	59,495
2006	42,708	38,648	58,094
2007	41,259	37,435	56,253
2008	37,423	34,172	50,660
2009	33,883	30,862	45,540
2010	32,999	30,296	44,862
2011	32,479	29,867	44,119
2012	33,782	31,006	45,960
2013	32,894	30,203	45,102
2014	32,744	30,056	44,950
2015	35,485	32,539	49,477
2016	37,806	34,748	52,714
2017	37,473	34,560	53,128
2018	36,835	33,919	52,286
2019	36,355	33,487	51,623
2020	38,824	35,766	54,272

It is too early to draw a correlation between a reduction in crashes and fatalities because of ADAS technologies being deployed. A deep dive into the data is needed to fully understand, and that is not the focus of this paper. However, it bears noting that the recommendation from researchers, OEMs, public sector deployers, private sector vendors and others in this space is to implement ADAS and ADS technologies as rapidly as possible.

Travel patterns and roadway usage will change over time just as they have for the last 100 years. Public sector agencies have a wealth of opportunity within their grasp, and harnessing this through a comprehensive innovation and transportation technology program is the key to success. This paper will detail readiness projects, cross-border collaboration, legislative action, outreach to the public, partnerships and other ways public sector agencies are preparing for this next generation of vehicles, drivers, and roadway architecture. The paper will focus on passenger vehicles with casual mention of some freight activities, as the inclusion of ADS for freight is an entirely different discussion. CAV for safety, mobility, advancing economic opportunity and workforce considerations as well as challenges to this preparation will also be discussed.

1.1 Background and Description of Non-technical Issues

1.1.1 Safety

As evidenced above by the NHTSA and IIHS statistics, there is an alarming trend of roadway fatalities in the United States. While uneducated public opinion may differ, preliminary deployments, some research and the promise of CAV technologies show that ADS features are safe and can contribute to fewer crashes and fatalities. Kutela et al. (2022) noted "The results indicate that an AV is at fault only for a small proportion of crashes (14.4%), and the likelihood of physical injury is about 2.8% higher when the automated vehicle equipped with ADS is at fault than the situation when a connected vehicle is at fault. This result implies the AV technology has to be precisely designed to reduce the likelihood of AVs being at fault to leverage its safety advantages." Kutela et al. used California crash records from January 2017 through October of 2020 where an AV was involved for this analysis and looked at three different interrelated variables: vehicle at fault, collision type and injury outcome.

1.1.2 Mobility

An aging transportation system that was designed based off of traffic modeling from 50 years ago can contribute to a loss in delay through bottlenecks, increased traffic volume, traffic incidents/crashes and a wide variety of vehicle and roadway geometry characteristics. In relation to safety, crowded roadways with inattentive drivers and higher speeds contribute to increased crash frequency and severity. A 2020 study (Shams et al., 2020) explored the contribution CVs must travel time reliability. Their study results "showed CV technology has the potential to reduce travel time due to speed harmonization and closer vehicle spacing. Under normal traffic conditions, increasing the penetration level of CVs up to 75% resulted in a higher average speed of cars. It seems, after 75% CV car platoon, made the opportunity for non-CVs to travel with higher speed in comparison to the platoon". Largely, connected vehicles require shorter following distances and are better informed of the traffic environment around them, leading to the likelihood of better mobility. The Shams study was a microsimulation model that tested different traffic patterns and variables, including different penetration rates of CVs.

Lin et al. (2016) notes "Impacts caused by congestion will be reduced, as CV and AVs will be able to drive closer together, increasing roadway capacity without impacting safety, since vehicles can keep minimum distances and still drive safely when compared with a human driver, due to their ability not accelerate and brake in unison."

1.1.3 Economic Opportunity

With few exceptions, the passenger vehicle form has remained unchanged for decades. An internal combustion engine with a steering wheel and brake pedal was standard until the emersion of electric vehicles and automated vehicles which introduced the possibility of designing and operating a vehicle without a human operator and/or passenger while leveraging the significant advances in battery technology. Legacy passenger vehicles received traveler information from Highway Advisory Radio (HAR) or point-specific Variable/Dynamic Message Signs (VMS/DMS) that could be viewed by the human operator. A small sector of the passenger vehicle fleet can now accept messages directly to the vehicle through various forms of connected vehicle technology such as Dynamic Short-Range Communication (DSRC), low latency communication or cellular communication offering real-time traveler information from a variety of public and private data sources, free or charge or available via subscription services. Some States are even changing the definition of a vehicle in their state statues. Alonso et al. (2022) notes "given the central role of mobility for our society and economy, the implications of a transformation in the transport sector will not be limited to transport but will regard many other aspects of our society." As the vehicle fleet shifts, new opportunities will abound – vehicle maintenance, programming and technology will enhance the quality of life for citizens. Ease of trips, whether discretionary or essential, will greatly improve the ability for citizens to get to and from places of work, community events and will improve access to healthcare by providing more options where a human driver and individual vehicles are less relied upon.

1.1.4 Workforce

Many OEMs have set and missed self-directed timelines for the deployment of driverless vehicles. A small percentage of the vehicle market today have levels 1–2 ADAS technology (NHTSA, 2022), however fully driverless vehicles are not yet widely deployed in the United States. Limited cities have fully automated vehicles available for public use and many additional cities have Level 4 AV shuttles deployed. A shift in worker roles is part of the transportation industry, it has just been more than 100 years since something as significant as the transition from horses to vehicles. The transition to CAV will likely modernize the workforce in many ways, but accommodations and resources are needed for education, outreach and for hands-on learning are paramount especially as ADA capable vehicles, not requiring a human operator, become more available. This moving target can make planning for future workforce considerations difficult. However, small shifts are being incorporated currently. "If we are to adequately prepare the workforce of tomorrow to meet the demands of the changing nature of work, we need to understand how jobs are expected to change over time. One of the challenges of understanding how jobs will change is having an accurate timeline for projected changes in specific industries. Changes due to automation in the transportation industry are one area that society must grasp if we are to prepare to meet the changing demands of the workforce of tomorrow" (Agrawal et al., 2023). Intentional planning and funding is needed to support worker transition and to obtain the support of the many labor unions throughout the transportation sector.

1.1.5 Outreach and Education

The allure of driverless vehicles has been the stuff of legend for decades. Flying cars and futuristic depiction of driverless cars in blockbuster movies can make outreach and education for actual availability of these technologies difficult. Outreach and education for these technologies are critically important, because if the traveling public does not understand or trust the technology, they will not support it at a policy level nor will they feel safe using the technology. There is a necessary blend of advocacy from the public and private sectors as well as a role for OEMs to play for outreach. "As the new era dawns on automated driving, it is important to delve into the aspects of consumer and user perception and the expected market penetration of this state-of-the-art technology. At the beginning stages, the public may have less confidence in automated vehicles, and drivers may resist forfeiting control of their cars. Successful demonstration projects are necessary to show the benefits and establish confidence in AVs to the public" (Lin & Wang, 2013). Public and private organizations must align messaging to guide the public's use and understanding of ADAS and ADS technologies that are fundamental to a CAV.

1.1.6 Readiness and Projects

Multi-state corridor coalitions focusing on traffic management and roadway operations have been in place for decades. As transportation priorities and travel patterns change, these coalitions have reprioritized work for CAV, which has resulted in projects like what is discussed in this paper. The Eastern Transportation Coalition (TETC), formerly the I-95 Corridor Coalition, started as an informal group of transportation professionals working together to manage highway incidents that impacted travel across state lines. In the past 25 years the Eastern Transportation Coalition has evolved from a small, highway-focused group to more than 200 public agencies working together to address the pressing challenges facing the eastern corridor.

During recent years, TETC has collaborated on several projects including work with Consumer Reports, the Connecticut Department of Transportation, and the University of Connecticut on a roadway readiness study. This study looked at the ability for vehicles with ADAS currently available on the market to understand and interpret pavement markings. In addition, TETC is currently working with JD Power on a driver training module to better understand customer sentiment and awareness of ADAS features. Again, collaboration is essential, and partnerships are needed with public and private sector entities as well as OEMs and advocacy groups. The ability to bring together all of these stakeholders is one benefit of a cross-border coalition like TETC. "Having a venue for engagement between the governmental agencies and the AV industry and having standards or best-practice guidelines related to infrastructure are very important for AV research and development," (Wang et al., 2022). Several public sector CAV pooled funds are focused on readiness studies. The New England Transportation Consortium (NETC) recently completed a project on cross-border policy and legislation readiness. The study

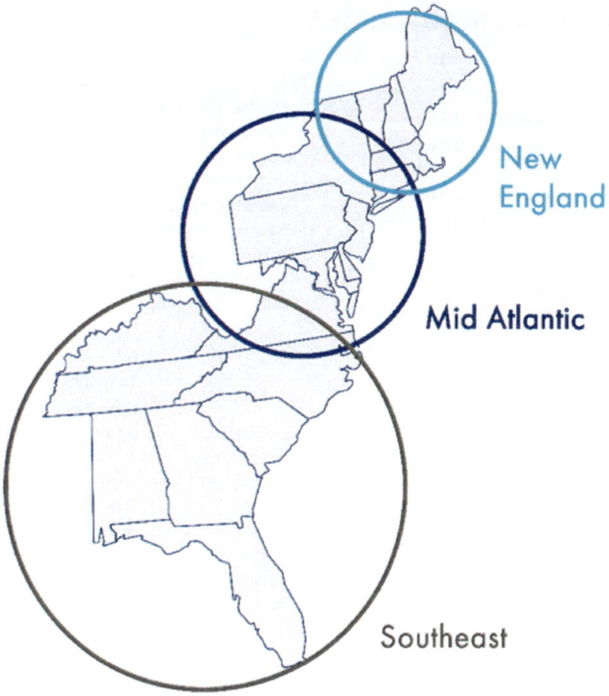

Fig. 1. The Eastern Transportation Coalition CAV Working Groups

answered important questions about cross-border collaboration and determined several key, meaningful next steps for CAV deployment and integration. A summary of these recommendations can be found in Fig. 2.

The readiness discussion would not be complete without a discussion of challenges. Many institutional knowledge silos exist, and proprietary data from vendors, start-ups and OEMs can make data sharing difficult. The lack of substantial federal guidance, and the stance of federal entities to remain technology agnostic can also present challenges. "Concerning data sharing, most companies are willing to share data to improve roadway infrastructure and increase AV safety," (Wang et al., 2022). Transparency in government regulations vary from state to state, but in general this necessary approach to an accountable government creates barriers to public private partnerships due to the inherent impacts to protecting intellectual and proprietary information.

2 Summary of the Discussion

This paper details some of the efforts to understand, influence, educate and integrate CAV technology. The ability to coordinate can occur at the local or regional/multi-state, national level. Each of these levels have an entirely different set of circumstances and should be carefully considered. The overall goal is to ensure a deployment landscape that is consistent enough for the public to understand and for public agencies to deploy.

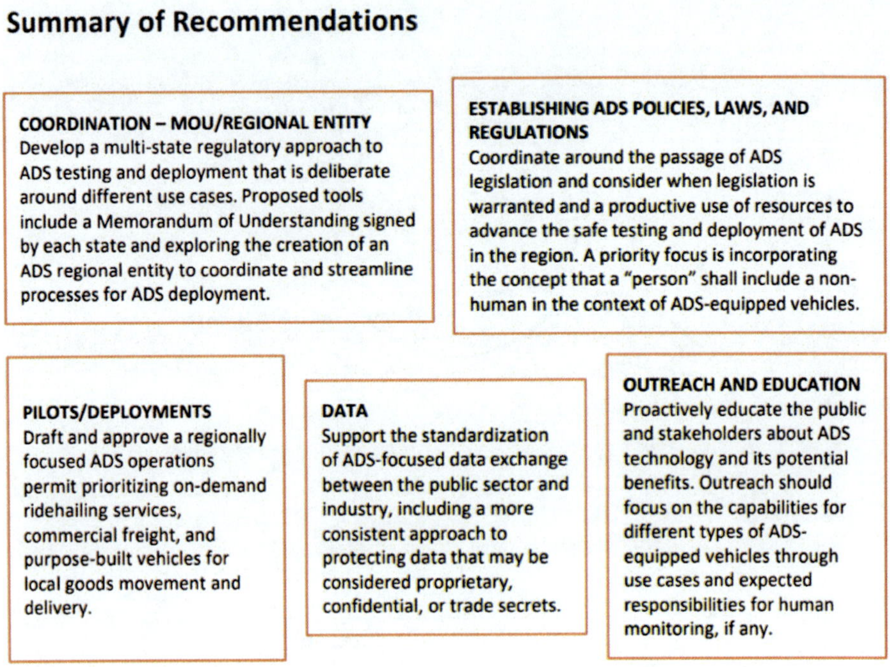

Fig. 2. New England Transportation Consortium (NETC) 20-4 Graphic

2.1 Context and Scope

Usman et al. (2020) states "Regarding the development of roadway infrastructure to support CAV operations, each stakeholder of CAV operations faces both challenges and opportunities. For multimodal road users (such as drivers, passengers, users of on-demand micro mobility, cyclists, and pedestrians), there are challenges (and opportunities) regarding how CAV-related infrastructure could influence their on-road travel efficiency, safety, and comfort level. Regarding technology and automobile companies, an immense challenge is to design vehicle connectivity and automation capabilities that are consistent with the capabilities of the road agency to accommodate these features. For highway agencies and policy makers, their association with the government presents an opportunity to leverage existing political structures to facilitate dialogue and reach consensus regarding infrastructure needs from the perspective of CAV policy. Regarding transportation planners, there will be challenges to ensure that CAV-related infrastructure is planned and designed in ways that promote planning paradigms including quality of life, accessibility, equity, and sustainability in a cost-effective manner." This is a succinct and complete assessment of the challenges and benefits of deploying at a local level. Who owns, operates, and maintains the CAV infrastructure, public or private, is also a central transitional discussion.

2.2 Recent Research and Trends

Regional collaboration can have many successes once the barriers are overcome. Funding, individual state/agency priorities, jurisdictional boundaries, facilitation, determining priorities and continual, ongoing maintenance needed to have beneficial regional collaboration can all be hurdles to adoption. The Eastern Transportation Coalition is a unique case study and model for how CAV can be advanced throughout a multi-state region. While TETC has 17 member states plus the District of Columbia Department of Transportation, there are also over 200 affiliate members – small municipalities, law enforcement agencies and metropolitan planning organizations. TETC began as an Intelligent Transportation Systems (ITS) priority corridor focusing on the Interstate 95 region more than 25 years ago. Since then, TETC has been rebranded and has transitioned to an entity fully funded by membership dues and some grants. This unique model allows the member states to drive the projects and priorities of the Coalition which encourages participation.

The Eastern Transportation Coalition (TETC) is an excellent example of the complexities of inter-state travel coordination and collaboration. The east coast of the United States has congested roadway networks with high volumes. Ease of travel from state to state for citizens could be challenging if the policies for how CAV are integrated differ greatly. The TETC CAV Working Group has actively engaged members from nearly every state on the east coast. While the specific priorities may differ between each of the three regional CAV working groups, the overall goals and objectives remain consistent.

Regional collaboration successes are visible throughout the United States. A 2018 study (Zhao et al.) about regional CAV deployments in the Austin area noted "Overall, at least two-thirds of Austin's auto users are forecast to elect an AV mode once those become widely available, either as a privately owned CAV or via fleets of SAVs." This shows an interest from consumers and citizens to utilize CAV technologies. Similar to the local level, regional/multi-state coordination must include Metropolitan Planning Organizations (MPOs), cities, counties, law enforcement as well as the state agencies in planning and deployment initiatives.

The Mid-America Association of State Transportation Officials (MAASTO) published a report with initiatives and priorities through 2030. The document shows short-medium- and long-range strategies for incorporating CAV including many of the components listed in this paper. There are many entities in the CAV space with a variety of interests. The CAV industry has a great amount of potential for new business which can create competitive silos of data and ideas. Many vendors are reluctant to share trade secrets and data which can lead to difficulty for state agencies when comparing data and equipment. The politics of CAV are emerging more every day. This is one of the reasons why public sector entities need sound guidance and increased data transparency for decision making (Fig. 3).

2.3 New Insights and Suggestions

Guidance and leadership from the national level is essential to avoid patchwork deployments. While the ultimate decision should be at the local/regional level, guidance from NHTSA, USDOT/FHWA, the American Association of State Highway Transportation

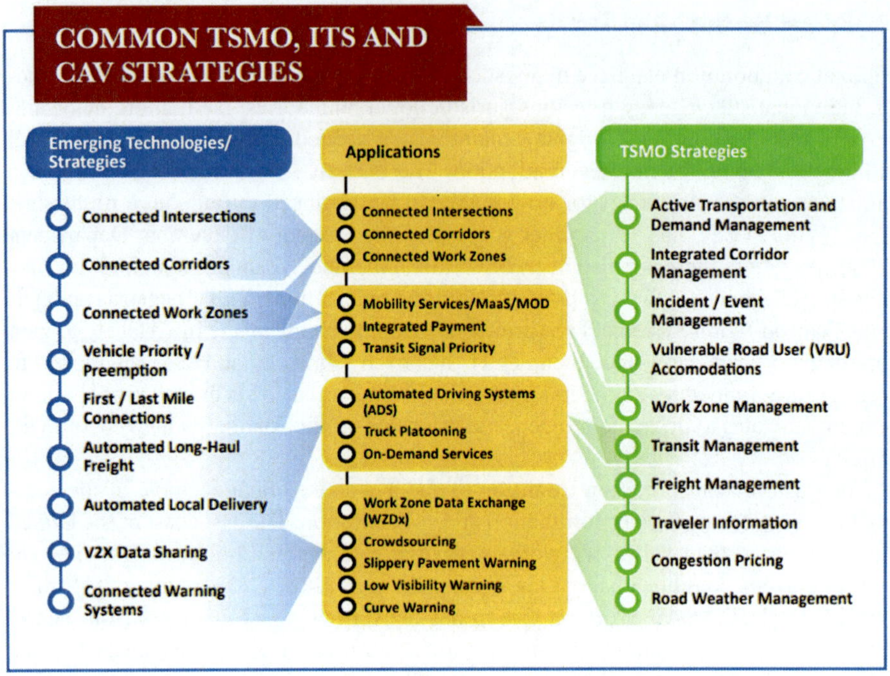

Fig. 3. MAASTO 2030 CAV Priorities

Officials (AASHTO), the Institute of Transportation Engineers (ITE), ITS America (ITSA), SAE, and other public and private trade organizations are important. For example, a 2022 Federal Register announcement from NHTSA seeks to standardize and provide guidance on occupant protection for vehicles equipped with ADAS technologies. The Register (2022) states "this final rule makes clear that, despite their innovative designs, vehicles with ADAS technology must continue to provide the same high levels of occupant protection that current passenger vehicles provide. This final rule updates the standards in a manner that clarifies existing terminology while avoiding unnecessary terminology, and, in doing so, resolves ambiguities in applying the standards to ADS-equipped vehicles without traditional manual controls." Standardization from this level can provide excellent guidance and baseline recommendations for local and regional level deployments.

3 Conclusions

Collaboration at all levels is a key to successful CAV planning, operations, and deployment. There are many considerations, some of which are underway and some of which are undetermined. A long-range strategy facilitated by public and private sector leadership organizations and trade associations are needed to ensure a cohesive role out of CAV near-term (ADAS) and long-term (ADS). The United States should continue to learn from deployments, strategies and frameworks in other countries/regions from around the globe.

As the transportation system ages, states will continue to rely on federal funding to make needed safety, mobility and multimodal improvements to their networks while shepherding the use of 3rd party data and leveraging public private partnerships. The transportation industry has learned a great deal through deployments, research and other testing in the last few decades regarding CAV technologies and deployments, however the rapid advancement of these technologies makes it somewhat difficult to stay current to differential near-term possibilities from long-term aspirations. Either way, the workforce of tomorrow needs to be trained in the emerging career paths that incorporate CAV. State agencies need to evolve to meet the changing needs of the transportation system, including retaining and retraining their workforces. One this is for certain, if properly harness and guided the incorporation of CAV technologies provides a promising opportunity to increase roadway safety, mobility and economic opportunity.

4 Next Steps

How will the discussion continue from the breakout session? Who should know about the outcome and how will the information be disseminated? Who else might need to get involved in the discussion? Should the scope of the discussion be reconsidered? Which research projects should be considered or reconsidered?

Acknowledgements. This paper was influenced by many mentors and practitioners across the United States as well as existing bodies of research. Within this profession, there are dozens of professionals who have dedicated their career to advancing technologies that will enhance roadway safety and mobility. These professionals openly and widely share their knowledge with the intention of informing and educating the next generation of transportation professionals. The contents reflect the views of the author and do not necessarily reflect the official views or policies of an employer or other agency.

References

Agrawal, S., Schuster, A., Britt, N., Mack, A., Cotten, S.: Building on the past to help prepare the workforce for the future with automated vehicles: a systematic review of automated passenger vehicle deployment timelines. Technol. Soc. **72**, 7–15 (2023)

Automated Vehicles for Safety. National Highway Traffic Safety Administration (2021). https://www.nhtsa.gov/technology-innovation/automated-vehicles-safety

Insurance Institute for Highway Safety (IIHS). Fatality Facts 2020: Yearly Snapshot (2022). https://www.iihs.org/topics/fatality-statistics/detail/yearly-snapshot

Kutela, B., Avelar, R.E., Bansai, P.: Modeling automated vehicle crashes with a focus on vehicle at-fault, collision type, and injury outcome. J. Transp. Eng. **148**(6), 1–11 (2022)

Lin, P., Wang, Z., Guo, R.: Impact of connected vehicles and autonomous vehicles on future transportation. Bridg. East West 46–53 (2016)

Mid-America Association of State Transportation Officials (MAASTO). 2030 CAV Regional Strategy (2020). https://www.maasto.net

National Highway Traffic Safety Administration. National Statistics (2021). https://www-fars.nhtsa.dot.gov/Main/index.aspx

National Highway Traffic Safety Administration. Newly Released Estimates Show Traffic Fatalities Reached a 16-Year High in 2021 (2022). https://www.nhtsa.gov/press-releases/early-estimate-2021-traffic-fatalities

Federal Register. Occupant Protection for Vehicles with Automated Driving Systems (2022). https://www.federalregister.gov/documents/2022/03/30/2022-05426/occupant-protection-for-vehicles-with-automated-driving-systems

Raposo, M.A., Grosso, M., Andromachi, M., Krause, J., Duboz, A., Ciuffo, B.: Res. Transp. Econ. **92**, 1–12 (2022)

Saeed, T.U., Alabi, B.N., Labi, S.: Preparing road infrastructure to accommodate connected and automated vehicles: system-level perspective. J. Infrastruct. Syst. **27**(1), 1–3 (2021)

Shams, A., Soltani, A., Niloofar, A., Famili, A.: The impact of connected vehicles on freeway travel time. In: International Conference on Transportation and Development, pp. 206–214 (2020)

Wang, P., McKeever, B., Chan, C.: Automated vehicle industry survey of transportation infrastructure needs. Transport. Res. Board. **2676**(7), 554–569 (2022)

Zhao, Y., Kockelman, K.M.: Anticipating the regional impacts of connected and automated vehicle travel in Austin, Texas. J. Urban Plan. Dev. **144**(4), 1–10 (2018)

Unscrambling the Automated Vehicle Policy Puzzle: AV Policy Development and Regulation in a New Normal

Katherine Kortum[1](✉) and Baruch Feigenbaum[2]

[1] Transportation Research Board, 500 5th Street NW, Washington, DC 20001, USA
kkortum@nas.edu

[2] Reason Foundation, 1747 Connecticut Avenue NW, Washington, DC 20009, USA
baruch.feigenbaum@reason.org

Abstract. An Automated Road Transportation Symposium (ARTS) breakout session entitled "Unscrambling the Automated Vehicle Policy Puzzle: AV Policy Development and Regulation under a New Normal" provided an overview of federal, state, and local policy. At the federal level, speakers discussed regulatory status, federal transportation policy points, and an industry prospective for federal action. During the second half of the breakout session, speakers provided perspectives and details about state and local efforts to design and implement AV systems that address societal goals, which include environmental benefits, reduced congestion, and improved mobility for all people.

Keywords: Automated vehicles · autonomous systems · federal policy · state policy · legislation · regulation · mobility · safety

1 Introduction

Disruption has defined our world over the last few years and continues to influence policy decision-making at all levels. Currently, major challenges in supply chains, public health, the economy, and environment have the attention of our public officials. Given this policy landscape, what is a desirable approach towards automated vehicle (AV) policy development and innovation?

We have also seen a once in a generation investment in the nation's transportation infrastructure through the Infrastructure Investment and Jobs Act, passed in 2021 [1]. How does this point to opportunities for bringing AVs into a future that supports improved outcomes in safety, equity, mobility, and the environment?

This session will focus on how AV policy-making fits into our 'new normal' and will include policymakers from all levels of government to tackle these important questions.

2 Presentations on Federal Policy

Presentations occurred in two parts. During the first half of the session, speakers focused on an overview of federal automated vehicle policy, discussing current approaches to developing the policies. During the second half, described in the next section of this

G. Meyer and S. Beiker (Eds.): ARTSymposium 2022, LNMOB, pp. 51–59, 2023.
https://doi.org/10.1007/978-3-031-34757-3_5

chapter, speakers from cities and states discussed how they are considering emerging automated vehicle policy topics, including equity, accessibility, and the environment.

2.1 A Status Update on Federal Rulemaking Projects Related to Automated Driving Systems

Marc Scribner, Senior Transportation Policy Analyst at the Reason Foundation, provided a status update on a variety of federal rulemaking projects. First, the *Final Rule on Occupant Protection for Vehicles with Automated Driving Systems (ADS)* was released on March 30, 2022, and was the first ADS-specific rule to be published by the federal government [2]. It was drafted to minimize changes to the Federal Motor Vehicle Safety Standards (FMVSS) while still preserving safety goals. This rule amends the crashworthiness (200 series) FMVSS to bring unconventional ADS-equipped vehicle designs into compliance. Notably, it clears the way for self-certification of occupant-less cargo vehicles.

Four ADS rulemaking projects are currently active and in the pipeline, three from the National Highway Traffic Safety Administration (NHTSA) and one from the Federal Motor Carrier Safety Administration (FMCSA). NHTSA's first project, *Considerations for Telltales, Indicators, and Warnings in Vehicles Equipped with Automated Driving Systems*, is estimated to be published in September 2022. A second from NHTSA is *Facilitating New Automated Driving System Vehicle Designs for Crash Avoidance Testing*; the comment period for this project closed on August 28, 2019, and NHTSA plans to begin analyzing comments by December 2022. NHTSA's third project, *Framework for Automated Driving Systems Safety*, had its comment period close on April 1, 2021, and NHTSA plans to begin analyzing comments by September 2022. The final active rulemaking project is from FMCSA, entitled *Safe Integration of Automated Driving Systems-Equipped Commercial Motor Vehicles*. Its publication estimate is January 2023.

In addition to the ADS-specific rulemaking projects, six ADS-adjacent rulemaking projects are also underway, five from NHTSA and one from FMCSA. FMCSA's project, *Automatic Emergency Braking Systems*, is estimated to be published in January 2023. All of the remaining ADS-adjacent projects are NHTSA's. The most near-term is *Expansion of Temporary Exemption Program to Domestic Manufacturers for Research, Demonstrations, and Other Purposes*, estimated to be published in October 2023. Two rulemakings are required by 2021's Infrastructure Investment and Jobs Act: *Light Vehicle Automatic Emergency Braking (AEB) with Pedestrian AEB* (estimated publication date of December 2022) and *Heavy Vehicle Automatic Emergency Braking* (estimated publication date of January 2023). A fourth NHTSA rulemaking is *Minimum Performance Standards for Lane Departure Warning and Lane-Keeping Assist Systems*, estimated to be published in February 2023. Finally, NHTSA also has *Alternative Options for Rearview Mirrors*. The Advanced Notice of Proposed Rulemaking comment period for this last project closed on December 9, 2019, and NHTSA plans to begin analyzing comments by December 2022.

It is also worth noting that under Transportation Secretary Buttigieg, three former ADS rulemaking projects and one former ADS-adjacent rulemaking project at NHTSA have been either withdrawn or terminated.

2.2 USDOT Perspective on Federal Policy and Initiatives

Vinn White is Senior Advisor for Innovation in the Office of the Secretary within the US Department of Transportation. His role includes coordinating Secretarial initiatives on emerging transportation technologies and building strategic partnerships across the Department to support the implementation of innovation-related opportunities.

Mr. White highlighted four points at the beginning of his presentation. First, transportation safety is facing a crisis that requires action. Second, in taking action, it is important to be thoughtful, to be deliberate, and to follow the science. The crisis is such that we cannot afford to create new challenges in pursuit of singular solutions. Third, the USDOT is working hard on transportation safety, guided as always by its principles. Fourth and finally, the USDOT cannot solve the crisis alone but requires partnerships.

The Department began approaching automation under Secretary LaHood and continued under Secretaries Foxx and Chao. Now a crisis exists on the roads, with NHTSA projecting that 43,000 people died on the roads in 2021, is a 10% increase over 2020 and a 20% increase from before the pandemic [3]. The Department's primary role as a safety regulator means that it must ensure all available tools are used to address these deaths; automation is one potential tool. USDOT is also using the rules and regulations described in the previous presentation, its discretionary funds, its ability to convene stakeholders, and its significant voice to call attention to these issues. If implemented correctly, AVs can make the system safer, improve Americans' quality of life and help to increase global competitiveness and create entirely new markets with quality jobs.

AVs can also enhance equity and address first- and last-mile options, helping to provide mobility to the mobility-challenged. In January of 2022, the USDOT released six innovation principles [4]: serve our policy priorities, help America win the 21st century, support workers, allow for experimentation and learn from failure, provide opportunities to collaborate, and be flexible and adapt as technology changes. In addition, the USDOT must work to foster integration, not just of external systems, but internally as well.

Equity and workforce issues are unlikely to improve without stakeholder action. The industry, public, USDOT, NHTSA, and regulatory agencies all play an important role in these issues. This is true not just for passenger vehicles but also in the realm of technology that can improve freight and transit and can combine with the built environment to increase safety. The potential implementation of each of these technologies is exciting, but often the messaging and media can get ahead of reality. The public needs clear information on what automated driving systems are and what they can and cannot do today.

Nothing in federal law precludes ADS-equipped vehicles from operating on public roads as long as they comply with both FMVSS and with state laws. In addressing these vehicles, the USDOT is taking a data-driven approach. Further modernizing FMVSS begins with data, and research will include system safety performance, human factors, and crashworthiness of novel and alternative seating positions. By now, most in the AV world are familiar with the Standing General Order on Crash Reporting for Level 2 vehicles, issued in June 2021, which requires reports of a crash of any ADS or Advanced Driver Assistance Systems (ADAS) equipped vehicle to be made within 24 hours [5]. NHTSA recently released data on incidents in the first year of this order. Further refinement can help the industry learn from this data, such as providing a better understanding of context and measure of exposure. Standing General Order reports are one useful data point, but are not a way to compare makes and models, nor whether humans or computers are better drivers. Instead, the USDOT can use them to flag individual events for investigation, providing the opportunity to track safety of systems. This tracking helps to identify risk models and improve the overall quality of available data.

The USDOT continues to work with vehicle manufacturers on best practices in the AV realm. A final rule on occupant protection is coming soon. The Department has also actively engaged AV developers for exemption petitions, with the first provided to Nuro in 2020 [6]. NHTSA's full regulatory portfolio is fairly ambitious, trying to identify all levels outside of its internal process that will help to achieve societal goals. Meanwhile, development in commercial motor vehicle automation is continuing, with rulemaking underway to update regulations around operations and maintenance for these vehicles. FMCSA is analyzing warning devices for stopped motor vehicles and in work zones. Updates to the Manual on Uniform Traffic Control Devices are underway, the Federal Transit Administration is looking for input on the Strategic Transit Automation Research program, and the USDOT Office of the Secretary is seeking input on design challenges launched under the previous administration.

The USDOT is taking a whole of government approach to deployment. In the age of innovation, the Infrastructure Investment and Jobs Act provides even more opportunity to engage with innovation. The Department is looking at opportunities for state, local, and tribal partners and will launch new opportunities soon for further engagement. The Secretary of Transportation and USDOT as a whole look forward to further partnerships.

2.3 Industry Perspectives on Federal AV Policy

Aravind Kalais, Advanced Technology Policy Director at the Volvo Group, provided an overview of one industry representative's perspective on AV policy. The Volvo Group is a global company that provides construction equipment, marine and industrial engines, trucks, and coach and transit buses. Volvo passenger cars are a separate company.

In 2020, Volvo Group consolidated its Society of Automotive Engineers (SAE) Level 4 automation initiatives throughout the company into Volvo Autonomous Solutions, one of only two new business entities in over twenty years. This consolidation occurred just before the pandemic began. Volvo Autonomous Solutions works both internally and with industry stakeholders on three pillars: confined area operations (mining, quarrying, ports), first mile (ports and logistics centers and a strategic development partnership with NVIDIA), and middle miles (hub-to-hub highway operations).

The Volvo Group believes that federal policies are needed to establish US leadership in automated vehicles. These policies should be part of a coordinated national strategy and vision to foster innovation and safe deployment of automated vehicles. Federal policymakers should discourage a patchwork quilt of state regulations and establish clarity between the roles of federal and state DOTs.

The USDOT should update and expand its suite of tools (AV 4.0, VSSA [Voluntary Safety Self Assessments], AV TEST [AV Transparency and Engagement for Safe Testing Initiative], and AV Hub) to inform future standards around AVs. The Department has taken an excellent approach but there are opportunities for more awareness.

The Department should also foster greater collaboration with the AV ecosystem to develop system-level functional requirements and compatibility testing. This is occurring to some degree, but there are many more opportunities for all parties involved to discuss what did and did not work. A Volvo truck needs to be able to react appropriately around any other Level 4 truck, and federal agencies have a role to help these discussions happen.

Federal policymakers can also promote transparency in AV real-world testing to address general public concerns. Media and company messaging exist, but it is important to know the parameters of a road test and the appropriate next steps.

The USDOT can and should modernize FMVSS to support AVs, and should identify and exercise existing USDOT statutory authorities for developing and manufacturing AVs in the US.

The federal government can fund digital and physical infrastructure to enable deployment of safe AVs. Connectivity is not a prerequisite for automation, but if the infrastructure owners and operators are moving toward sophisticated digital infrastructure to enhance safety, all parties would benefit. Volvo's trucks move from coast to coast and the company wants all aspects of the travel to work seamlessly without guesswork.

Finally, USDOT should support research and deployment initiatives to enable better understanding around AVs and should support technology investments that advance resiliency goals, sustainability, and mitigate climate change.

3 Presentations on State and Local Policy

3.1 Trenton MOVES

Alain Kornhauser is a Professor of Operations Research and Financial Engineering, along with the Director of the Program in Transportation at Princeton University. He is involved with Trenton MOVES, a transportation initiative to bring AVs into the city of Trenton, NJ [7].

Automation has to be safe and has to be equitable. Trenton MOVES is focused on equity, both in the initial deployment city of Trenton and also in the planned second city of Perth Amboy, NJ. In both cities, 70% of households have access to zero or one car. Mode split is proportional to auto ownership, meaning that 70% of the households have limited mobility. The inability to drive oneself around makes mobility difficult.

It is also important for mobility to be affordable, which is where technology comes into play. The only way to make transit more affordable is to not have an attendant. Nonetheless, to be high quality, the transportation service has to be on demand with easy access to all people. The goal of Trenton MOVES is to eventually blanket New Jersey with kiosks that serve as the onboarding and departure points of the service, analogous to the entrance of an elevator bank. The first fifty kiosks in the Trenton network have a maximum walk time of five minutes. They are intended to be in the community, designed by the community, and used by the community.

Success in Trenton would lead to an extension throughout Mercer County and then replication in other cities. The end goal is for the system to be one which even those who own their own vehicles want to use.

3.2 Improving Equity, Accessibility, and Environment Through Autonomous Vehicle Policy

Katie Stevens is the Head of State and Local Policy for the western region of Nuro, working with state and local leaders, agencies, and organizations to support electric and autonomous goods movement. The Nuro vehicle is small and will never hold a person. Instead, it has two doors that open to hold goods only. The company employs over 1,300 people located primarily in Houston and the Bay Area. Some testing occurs in the Los Angeles area, and the company also has a teleoperations center in Arizona and a manufacturing center in Nevada. It has delivery partnerships with 7–11, Kroger, and CVS, and was the first automated vehicle manufacturer to receive an exemption from certain NHTSA low-speed vehicle standard requirements [6]. This exemption applies to the R2 vehicle, which is no longer required to include features such as mirrors and a windshield.

3.3 AV Equity, Mobility, and Sustainability in Ohio

Cynthia Jones is the Project Manager for DriveOhio, which is the state's center for innovation, incubation, and implementation of autonomous and connected vehicle technologies. It exists outside of the Ohio Department of Transportation and is tasked with working with all of state government, a commitment it has had through multiple administrations. DriveOhio works both on the ground and in the air on all things connected, automated, and electric, and has had a great deal of its efforts adopted as a result of Columbus's Smart City grant in 2016.

Many such efforts are on equity issues. One program focuses on workforce training, making it available and allowing employee credentials so that companies can be reimbursed for their employees' credentials.

DriveOhio has a significant rural emphasis, given that 97% of the land mass in the United States is rural. Most AV testing is in the urban environment, but the organization is trying to broaden that testing experience. With the Infrastructure Investment and Jobs Act and its electric vehicle funding, a great deal of work will occur on electric charging in diverse ways. DriveOhio is also working to support those who want to be able to charge a vehicle but who are in apartments or otherwise do not have garages.

As part of its automated driving systems projects, DriveOhio is undertaking ten workshops later in 2022, all around the state, with those who are generally classified as vulnerable road users. The organization is seeking their feedback on what an ideal AV world looks like. In rural settings, many people choose to age in place. When residents of rural communities lose their ability to get around as they age, DriveOhio wants to be able to support them.

3.4 Equity and Mobility Through Nonprofit Community Based Volunteer Transportation

Katherine Freund is the founder and President of the Independent Transportation Network of America. ITN America delivered its first ride in June of 1995 and has been operating as a national nonprofit for 27 years. The company developed a model that is community-based, community-supported, and does not rely on taxpayer support in order to be financially sustainable. Instead, it has developed a variety of ways to pay for transportation services; programs exist in which people can trade their cars and volunteer drivers receive electronically-stored transportation credit in an e-wallet.

This model operates in many communities across the country, and ITN America is now taking the technology supporting it to Salesforce so that the technology can be available to small communities across the country at an affordable price. Transportation is expensive, and more than fifty million people in the country have no transit available to them at all, at any price. Twenty percent of the population lives in the 97% of the country that is rural [8], most of which is a transit desert. ITN America intends to begin by working with these people who have no transportation services.

Other technology endeavors the company is undertaking include Rides Inside, the largest database of senior transportation in the United States. It is searchable and available to the public, although it was developed with private resources. ITN America's work is supported by government, private philanthropy, and private industry. Its software's routing algorithms were donated by ESRI. Government, philanthropy, and private industry will all need to work together to make transportation sustainable.

3.5 Audience Discussion

After each state and/or local presenter spoke to the group, the audience discussed the presentation. The following are summaries of major points brought up with all in attendance.

Many governments are focused on sustainability issues, leading them to want to deploy AVs in a way that at least does not hinder, even if it does not outright help, the goals of reducing emissions and congestion. The status quo is certainly not working, with inequities widespread. Most AV companies assume that most, if not all, of their vehicles will be electric; Nuro, for example, manufacturers its vehicles as all-electric. Ideas for state and local governments include incentives to transition to electric vehicles across the modal spectrum, repurposing curbside parking for charging infrastructure, zero emission delivery zones such as in Santa Monica, CA, and support of safe street design that encourages the use of roads by all people and vehicles, including those that require little or no fossil fuel.

Equity has many definitions, but states and localities need to take certain steps to ensure equity as they deploy AVs. Service must be provided in places where people are currently lacking mobility. Understanding all transportation funding sources (62 within Ohio alone, for example) is a challenge, and research is underway to coordinate these services more effectively. Residents who are aging often want to stay in their own homes, and providing them the means to do so is both an opportunity and a challenge.

Deploying AVs will ideally increase access and allow transportation providers to target or reach communities who do not currently have good transportation access. In order to do so, we must think beyond transit agencies. ITN America's community-based approach moves people using the capacity that is currently parked in driveways. Governance will also make a difference; currently nonprofits are doing a great deal of work but they do not have the resources or the reach of government services. Autonomous delivery vehicles can provide benefits in paratransit and food desert areas, helping to reduce the cost of delivery to serve trips that others can't or won't. Bringing goods to people also provides an alternative to high-cost paratransit trips.

Automating any portion of the transportation system is likely to have effects on labor and workforce issues as well. Opportunities exist to design curricula, in coordination with community colleges and others, to better prepare people for upcoming changes. Nuro works with De Anza College in Silicon Valley on a two-phase certificate-based program that includes not only basic automotive technology courses but also introductions to Python and Linux programming languages. This certification leads to a junior fleet technician position at Nuro. AV shuttle pilot programs such as Trenton MOVES can complement transit by providing additional rides to train stations and other transit services. Many industries around Ohio, including trucking, have had trouble finding enough workers. Similar to Nuro, DriveOhio has begun capstone courses and other training efforts with universities around the state to create the workforce needed in the future.

4 Conclusions and Next Steps

Discussions about policy, whether at the federal, state, or local level, can be wide-ranging, and this breakout session was no exception. Presenters and the audience all agreed that both public and private sectors need greater certainty in the regulatory climate. In addition, both industry and the government can be the barriers to implementation; neither side has a lock on being the limiting party. Financial resources, however, are rarely the biggest problem in implementation. Projects that include equity, accessibility, and environmental components are most popular politically but can also be the most challenging to implement.

Multiple research questions emerged from the discussion. In the AV transition phase, how do we develop a roadmap for regulatory agencies? Is there an equivalent of an energy star rating for AVs, signifying that the vehicle meets certain standards from a professional body? And how do we develop and pilot a multi-state environment with a harmonized regulatory structure such as the Eastern Transportation Coalition?

Many questions remain to be answered, and many more questions will emerge in coming years. Continued discussions such as these can help make AV visions become reality and the research efforts suggested here can help to drive progress year after year.

5 Glossary of Acronyms

ADAS	Advanced Driver Assistance Systems
ADS	Automated Driving System
AEB	Automatic Emergency Braking
AV	Automated Vehicle
AV TEST	Automated Vehicle Transparency and Engagement for Safe Testing Initiative
FMCSA	Federal Motor Carrier Safety Administration
FMVSS	Federal Motor Vehicle Safety Standards
NHTSA	National Highway Traffic Safety Administration
SAE	Society of Automotive Engineers
VSSA	Voluntary Safety Self-Assessment

References

1. H.R. 3684 – Infrastructure Investment and Jobs Act. https://www.congress.gov/bill/117th-congress/house-bill/3684/text
2. Occupant Protection for Vehicles with Automated Driving Systems. 87 Federal Register 18560 (2022). https://www.federalregister.gov/documents/2022/03/30/2022-05426/occupant-protection-for-vehicles-with-automated-driving-systems
3. Newly Released Estimates Show Traffic Fatalities Reached a 16-Year High in 2021. National Highway Traffic Safety Administration, USDOT. (2022). https://www.nhtsa.gov/press-releases/early-estimate-2021-traffic-fatalities
4. USDOT Innovation Principles. United States Department of Transportation. https://www.transportation.gov/priorities/transformation/us-dot-innovation-principles
5. Standing General Order on Crash Reporting for Incidents Involving ADS and Level 2 ADS. National Highway Traffic Safety Administration, USDOT. https://www.nhtsa.gov/laws-regulations/standing-general-order-crash-reporting
6. Docket No. NHTSA-2019-0017. Nuro, Inc.: Grant of Temporary Exemption for a Low-Speed Vehicle with an Automated Driving System. https://www.nhtsa.gov/sites/nhtsa.gov/files/documents/nuro_grant_notice_final-unofficial.pdf
7. Trenton Moves. Corporation for Automated Road Transportation Safety. https://www.cartsmobility.com/trenton-moves
8. One in Five Americans Live in Rural Areas. United States Census Bureau. https://www.census.gov/library/stories/2017/08/rural-america.html

Paradigm Shift Beyond Business-As-Usual for Automated Road Transportation to Contribute to Climate-Neutral Smart Cities

Jaap Vreeswijk[1](✉), Danielle Chou[2], Wolfgang Backhaus[3], Isabel Wilmink[4], and William Riggs[5]

[1] MAP Traffic Management, Van Deventerlaan 20, 3528 Utrecht, The Netherlands
jaapvreeswijk@gmail.com
[2] US Department of Transportation, Turner Fairbank Highway Research Center, 6300 Georgetown Pike, McLean, VA 22101, USA
danielle.chou@dot.gov
[3] Rupprecht Consult, Clever Street 13-15, 50668 Cologne, Germany
w.backhaus@rupprecht-consult.eu
[4] TNO, Anna Van Buerenplein 1, 2595DA The Hague, The Netherlands
isabel.wilmink@tno.nl
[5] University of San Francisco, 2130 Fulton Street, San Francisco, USA
wriggs@usfca.edu

Abstract. One of the major challenges for society today is to reach environmental goals. Around the world governments aim to reduce transportation emissions and are investing in a diverse set of strategies, including connected cooperative and automated mobility. Many discussions about automated road transportation technologies revolve around automation feasibility, improving safety and improving the efficiency of transport networks, but we have not fully explored the potential of automation as a catalyst or enabler to transform society to an ecologically sustainable state. This chapter discusses ways that connected and automated technologies can address climate concerns as addressed at the ARTS22 workshop on automated road technologies and climate change.

Keywords: Autonomous Vehicles · Climate Change · Energy Use · Urban Policy · European Commission

1 Introduction

Cities are significant drivers of greenhouse gas (GHG) emissions, and the transport sector is a major contributor to these emissions. According to the European Environment Agency (EEA), the transport sector is responsible for about a quarter of total GHG emissions in the European Union (EU) [1]. In urban areas, transport emissions can account for an even higher percentage due to the higher population density and congestion as well as commuting to and from cities – with a high share of individual car usage.

G. Meyer and S. Beiker (Eds.): ARTSymposium 2022, LNMOB, pp. 60–71, 2023.
https://doi.org/10.1007/978-3-031-34757-3_6

Road transport is the primary contributor to transport emissions in urban areas, including cars, buses, and trucks. The European Commission estimates that road transport accounts for about 70% of transport emissions in urban areas. Cooperative, Connected and Automated Mobility (CCAM) promises advantages for the mobility of the future in terms of safety, accessibility, improved use of road space and also for the decarbonization of transport. Research has shown that this can be achieved through increasing the shared nature of automobiles, changing fuel types and business models, as well as increasing the performance of urban networks from network management standpoint [2-6]. Yet, most European local authorities have a lot of uncertainties about the introduction of CCAM on their road network and are missing guidance on how they can ensure its alignment with their policy goals. Many cities question whether CCAM implementation will fulfil the above-mentioned promises.

CCAM will have significant impacts on most transport and urban planning related activities of a city analyzing how it could affect road safety, traffic efficiency, infrastructure, socioeconomic aspects, travel behavior and spatial planning. Still, there is a high degree of ambiguity surrounding the assessment of these impacts as nobody, at this stage, can really predict how the technology will be used and whether the positive aspects will out-weight the negative ones. Without good preparation and planning, it could actually amplify the urban mobility problems that cities are currently already facing, and lead towards increases in the number of travelled kilometers, urban sprawl and congestion levels [7].

This uncertainty about the impact of CCAM on the decarbonization of transport is also reflected in the European Union's wide-ranging survey (see Fig. 1), which includes more than 27,000 respondents across Europe.

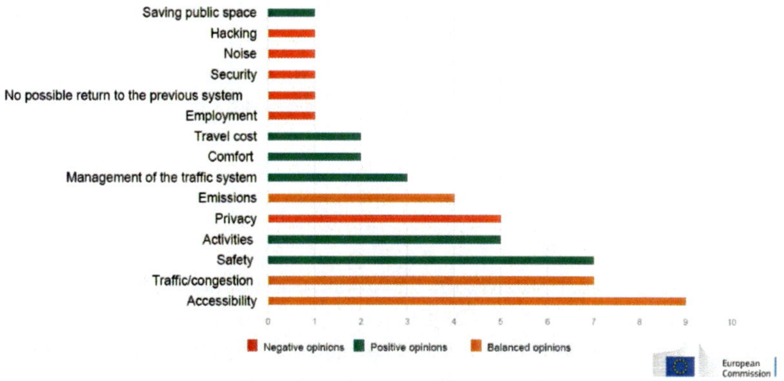

Fig. 1. Expectations and Concerns from a Connected and Automated Mobility [8]

As part of the Horizon Europe program, the EU has launched a Mission "100 Climate-Neutral and Smart Cities by 2030" [9]. The objectives of the mission are "to achieve 100 climate-neutral and smart European cities by 2030 and to ensure that these cities act as experimentation and innovation hubs to enable all European cities to follow suit by 2050". The European is that CCAM contributes to this goal of net zero emissions, by creating a mobility-oriented rather than driver oriented transport system. This includes better integration of CCAM in existing transit services and Mobility-as-a-Service application, and displacing car trips with other modes.

Simply employing CCAM technologies in a "business as usual" way with a focus on drivers and cars would likely be insufficient to truly address climate goals, thus a change in how we look at CCAM needs to come about. Automated vehicles in any way, shape or form are a means to an end, and in this case the end is to transition to a mobility system that is climate-neutral, and contributes to the well-being of people. The question is thus: how can automated vehicles, or how can connected, cooperative and automated mobility, become part of the (urban) climate agenda, the EU's carbon-neutral mission, and the new topic guide for sustainable urban mobility planning?

2 Examples of Why a Paradigm Shift is Needed

In the broader context numerous studies have dialogued the potential pros and cons of AVs. According to several research studies made in European research projects [10–12] CCAM has the potential to support the decarbonization of urban mobility systems in several ways:

- Energy-efficient driving: Connected and automated vehicles can be programmed to drive more efficiently, such as by accelerating and decelerating smoothly and avoiding unnecessary idling. This can result in significant fuel savings and emissions reductions.
- Alternative powertrains: CCAM technologies can also facilitate the adoption of alternative powertrains such as electric, hybrid, or fuel cell vehicles. Automated vehicles can be charged or refueled automatically, and their energy consumption can be optimized through intelligent routing and scheduling.
- Modal shift: CCAM technologies can also facilitate the use of alternative modes of transportation such as shared mobility, public transit, or active transportation (e.g. cycling, walking). This can reduce the number of single-occupancy vehicles on the road and reduce emissions.
- Traffic management: Connected and automated vehicles can communicate with each other and with traffic management systems to optimize traffic flow, reduce congestion, and reduce emissions from idling vehicles.

Thus overall, CCAM can support the decarbonization of urban mobility systems by promoting energy efficiency, facilitating the adoption of alternative powertrains, promoting modal shift, and optimizing traffic management. However, several studies have over the past years have raised concerns about the consequences of a "business as usual" approach towards CCAM. Research has indicated that with absence of policy the technology could significantly impact cities, their infrastructure, and their residents. For example, [13] reported on effects of AVs found with model calculations using an innovative transport model capable of modelling various new private and shared transport modes. Several future scenarios with L3/4 and L5 automated mobility concepts were explored, showing that if (electric) automated cars are available that are easy to use, cheap to operate and in the case of L5 vehicles do not require a driver, this would likely result in a shift towards higher use of private or shared cars ('robotaxis'), with people switching from walking, cycling and public transport to car modes. Also, with car km's travelled on the rise, the amount of delay increases - a strong mix of interventions is needed to keep delays at the same level as in the reference scenario.

A study for Ruter (the Oslo Region public transport company) looking at how autonomous vehicles may change transport in Oslo [14] shows the potential effects in different futuristic scenarios for the Oslo region. A transport model was developed to analyze consequences of autonomous cars and Mobility-as-a-Service (MaaS) systems. The results showed a range of effects. In the best case, traffic could be reduced by 14% (through possibilities for ride sharing), but in the worst case traffic volumes double which results in a complete traffic breakdown. In all scenarios, the number of cars used within the mobility system can be drastically reduced (up to a 93% reduction). Based on the model calculations, the study concluded that traditional public transport combined with cycling and walking will be key for future urban mobility concepts, as a mobility system solely relying on autonomous vehicles in a MaaS concept to cater for all transportation needs is not desirable, if they are used as cars are today.

The Oslo study followed the Lisbon studies carried out by the International Transport Forum team at the OECD (four studies in total, see e.g. [15, 16]), and also references studies in Helsinki, Dublin, Auckland and Stuttgart. All studies showed that in the most 'sharing-friendly' scenario, substantial reductions in the number of vehicles needed and vehicle km's driven can be achieved. However, in scenarios without ridesharing the amount of km's driven can increase (very) strongly, which can have negative implications for traffic safety and the environment. Typically however, the studies into impacts of AVs from before 2020 look at how AVs impact the performance of the transport system, not looking at the wider societal impacts on e.g. traffic safety, the living environment and energy use or public health. Accessibility is still the main concern (with an implicit link to how this contributes to economic growth). This is starting to change. The last Lisbon study [16] looks beyond accessibility with the question whether cities should make room for ride services at the curb. The report states that in most cities, the answer is yes, but that cities should base such changes on a "broader strategic re-assessment of the priorities regarding access and use shared public assets, including streets and curbs, they wish to give to different modes".

So what can be derived from literature is that AVs could impact traffic congestion and increase km's driven in cities due to increased demand for individualized transportation. While some of these studies using rideshare vehicles as a proxy have shown potential for increase in travel, more realistic studies in actual AVs show that service can be complimentary with existing transit and enhance access for those with reduced mobility and historically underserved populations [17–19]. In addition, other research has shown that AVs may have more longitudinal impacts on travel patterns related to urban form and land use in cities [20]. The increased efficiency of AVs may lead to changes in the spatial distribution of economic activity and residential development, as individuals and businesses may be willing to locate further away from city centers [21, 22]. While this could lead to changes in travel patterns and increased demand for road infrastructure in suburban and rural areas, research is uncertain and predicated on more complicated urban planning decisions [23].

Thus, these positive effects on decarbonization of transport have not yet been clearly demonstrated and there is a risk of rebound effects that may partially offset the potential decarbonization benefits. In summary these are:

- Induced travel: One of the main risks associated with CCAM is that it could lead to induced travel. This means that people may start using vehicles more frequently or over longer distances, which could increase overall energy consumption and carbon emissions.
- Empty vehicle trips: CCAM systems can improve vehicle routing and reduce the number of empty trips, but it may also encourage more trips where vehicles travel without passengers. This would result in higher energy consumption and emissions.
- Vehicle size and weight: CCAM systems may also encourage the use of larger and heavier vehicles due to the perception that these vehicles are safer or more comfortable. However, larger and heavier vehicles typically require more energy to operate and therefore emit more carbon.
- Technological lock-in: The widespread adoption of CCAM technology may lead to a technological lock-in effect, where the focus shifts away from other potential decarbonization solutions. This could result in a slower transition to other low-carbon transportation options.

2.1 Welfare Beyond GDP as a Starting Point for CCAM

The topic of welfare beyond GDP, or wellbeing economy, has gained interested over the past few years. In the Netherlands, several studies elaborated the concept for the mobility system. Figure 2 shows how there are several dimensions to welfare beyond GDP in mobility: impacts of the mobility system on the living environment, safety, accessibility and health. Also, based on the goals for sustainable development, it is relevant to look at distributional effects, taking into account various population groups, regions or time periods ("well-being 'here and now', 'later' and 'elsewhere'"; see [24]). And, in order to be able to explain the impacts of interventions in the mobility systems on the various dimensions, it is also necessary to evaluate the performance of the mobility system (with well-known indicators such as modal split, km's travelled, travel times, delays, queue lengths etc.).

When thinking about how automated vehicles or CCAM can contribute to reaching environmental or other societal goals, it helps to consider all four dimensions, distributional effects and indicators (all of which can also be split up for different population groups, or regions, and determined for different time periods) to see which impacts the AV- or CCAM-based interventions may have and which balance in impacts might be desirable.

2.2 Towards Sustainable Urban Mobility Planning

To minimize these potential rebound effects, it is essential to design CCAM systems in a way that encourages energy efficiency and low-carbon travel modes, such as walking, cycling, and public transportation. Policies and regulations can also be put in place to incentivize the use of low-emission vehicles and discourage unnecessary travel (Fig. 3).

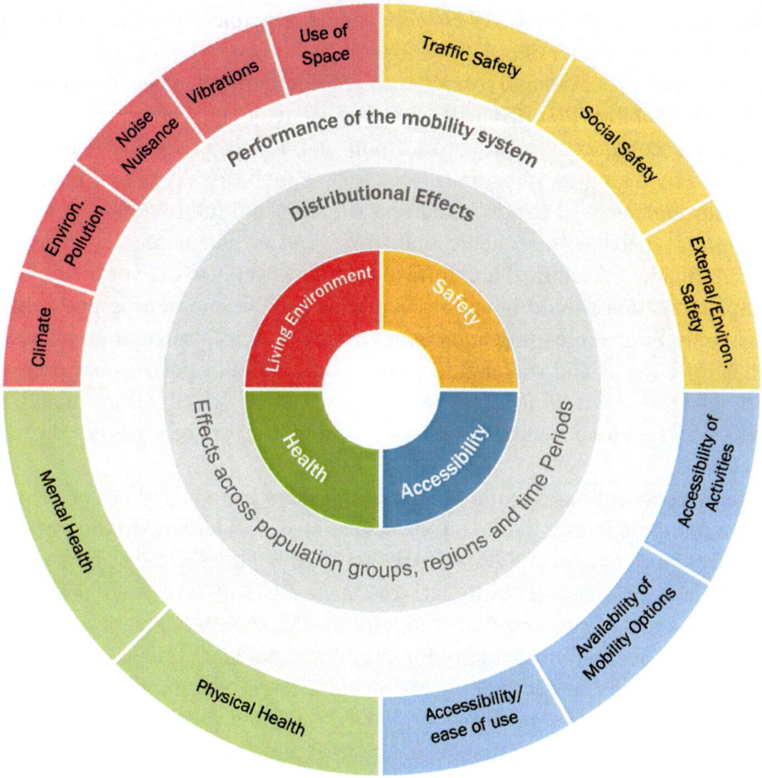

Fig. 2. Dimensions and subdimensions of welfare beyond GDP, and mobility [25].

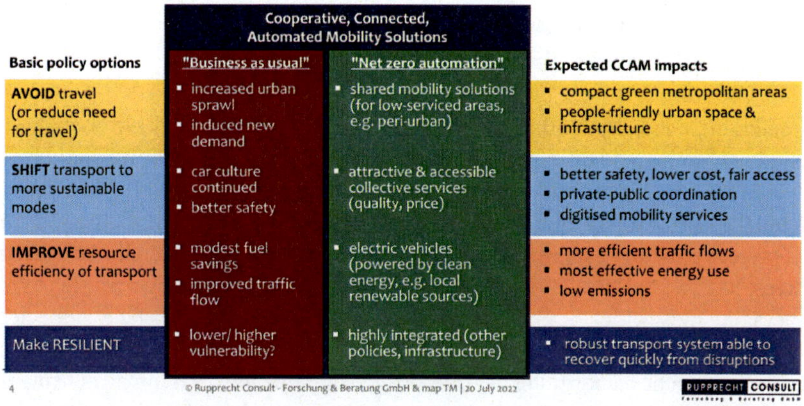

Fig. 3. Potential pathways of CCAM and impacts on transport decarbonization [26].

A strategic approach in planning a wide range of measures that will ensure a sustainable and just deployment of CCAM, and by this supports higher level mobility goals, can be achieved by following the SUMP concept and its principles. The European Commission has developed a number of policies and tools to facilitate the planning of sustainable urban mobility systems. First and foremost, the SUMP concept. The concept of the Sustainable Urban Mobility Plan (SUMP) [27] is a strategic and integrated approach for dealing with the complexity of urban transport and advocates for fact-based decision making guided by a long-term vision for sustainable mobility. As key components, this requires a thorough assessment of the current situation and future trends, a widely supported common vision with strategic objectives, and an integrated set of regulatory, promotional, financial, technical and infrastructure measures to deliver the objectives – whose implementation should be accompanied by reliable monitoring and evaluation. In contrast to traditional planning approaches, SUMP places particular emphasis on the involvement of citizens and stakeholders, the coordination of policies between sectors (transport, land use, environment, economic development, social policy, health, safety, energy, etc.), and a broad cooperation across different layers of government and with private actors.

The SUMP concept has become mainstream across Europe and is a centerpiece of European mobility policies, e.g. the European Commission's Urban Mobility Framework [28], addressing the concept of SUMPs to reduce road fatalities in cities, address climate change, and make use of new mobility services, such as CCAM. As part of the continuous updating process of the European Guidelines for SUMP, several topic guides for different policy areas have been published, e.g. for environmental and climate change planning (SUMP decarbonization guide [29]) or the integration of new mobility modes such as CCAM [7]. The aim of this guide for automated road transport planning is to provide guidance along the eight core principles of a SUMP, and how these can be applied in the context of CCAM, and best practice examples in order to increase the capacity of cities to introduce CCAM into their sustainable urban mobility planning processes.

3 Summary of Dialogue at ARTS22 Workshop on Automated Road Technologies and Climate Change

The session included about 30 attendees from primarily the US and Europe who were interested in learning and discussing how automation and/or connectivity could be judiciously applied to achieve climate neutrality goals. The session kicked off with presentation about the European Commission's (EC), approach to CCAM. The dialogue outlined the EC project portfolio specifically dedicated to achieving Europe's climate neutrality objectives, with a goal of making at least 100 European cities climate neutral by 2030 and leveraging these cities' lessons learned to help all European cities be climate neutral by 2050.

As outlined in the background the European vision for CCAM is aimed at using these new technologies - such as connected and automated vehicles (CAVs), micro-mobility, and MaaS – to create a mobility-oriented rather than a driver-oriented transportation system. This includes better integration with existing transit services and displacing car trips with other modes. The CCAM portfolio has a budget of one billion euros (500 million from the EC itself).

Subsequent presentations dove into greater detail about the types of impacts that connected and automated road transportation could have on safety and climate goals, with a pointed note that simply employing these technologies in a "business as usual" way with a focus on drivers and cars would likely be insufficient to truly address climate goals, and other technology and policy approaches would be more effective and available sooner. Yet, Europe continues to pursue these technologies because they have the potential to be transformative by upending public notions of how to travel most conveniently and cost-effectively. The hypothesis is that European approach is holistic and examines how multiple technologies and policy çan synergistically work to achieve climate neutrality and how such work can be done incrementally to make progress on an earlier time horizon.

3.1 Workshop Setup

Based on the presentations, the workshop then shifted to an interactive Design Thinking (DT) format where attendees used a virtual Google Jamboard to idea new concepts. Each table was assigned a problem space of either passenger or freight in urban, suburban, or rural areas. The session began with an icebreaker that asked each team to collectively create a piece of art (in this case, Andy Warhol's painting of Marilyn Monroe) using Jamboard tools in order to familiarize themselves with the tools. The teams were next guided through rapid brainstorming sessions on four questions:

- What is needed for transportation fueling?
- How can automation and/or connectivity help with fueling
- What is needed for transportation resilience?
- How can automation and/or connectivity help with resilience?

At the conclusion of this brainstorming, each team selected one concept from the brainstorming questions to mentally prototype for the next twenty minutes. Afterwards, each team presented its designs and ideas before closing out the DT activity.

3.2 Discission Insights and Suggestions

Each team's design insights and focus areas was different. One team stressed policies rather than technological solutions and discussed how AV deployments should focus on transit and freight rather than individual passenger transport. It expressed some skepticism that automated mobility itself would be effective at achieving climate neutrality. Rather, this team produced the novel insight that, instead of using connected and automated technologies for moving things, those same technologies could be applied towards enforcement of climate-friendly policies. This shift in perspective and framing was very unique and has been little explored by the research community, beyond the use of congestion and cordon pricing.

Another team, assigned to suburban freight, highlighted the fact that the term suburban is ambiguously defined. They eventually settled on identifying ways in which private freight vehicles could be called upon to join an emergency response. Such vehicles could enable more sharing during disaster or high demand periods to improve resilience. The team honed-in on the types of capabilities required for emergency response, including medical emergency equipment, refrigeration, and power generation; they observed the benefits of having flexible vehicle configurations and identified vehicle right-sizing as a critical parameter.

A third team designed a fleet of automated mobile charging stations to improve rural passenger travel. Charging infrastructure could thus more flexibly serve its dispersed constituents and could require less dedicated fixed infrastructure to serve a wide area. A snapshot of their prototype is shown in Fig. 4.

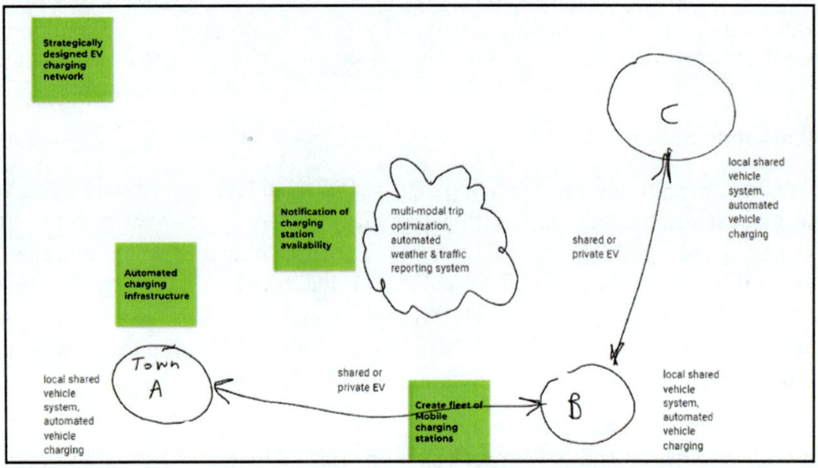

Fig. 4. Example of a potential strategically designed charging network.

The fourth team explored how connected and automated technologies could be applied to deal with weather and emergency response, including automated snowplows, automated emergency response, and automated heating elements (unclear where these would be applied) to deal with winter weather such as snow, ice, sleet, etc.

All in all, as shown in Fig. 5. The workshop successfully spurred its participants to re-think how connected and automated technologies might serve as elements of a "fifth fuel," energy efficiency and conservation. Particularly of note, were discussions as to how electric AV fleets could enhance load balancing and prioritize energy efficiency from a people-miles traveled standpoint, while increasing transportation access for those who have not been service priorities for public transport agencies in the past.

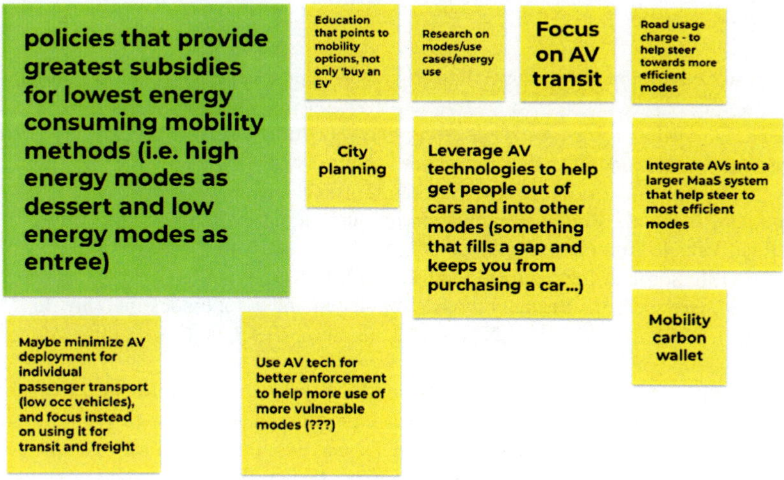

Fig. 5. Ideation of how AVs can prioritize low energy / carbon solutions.

4 Conclusions and Next Steps

In conclusion, AVs inspire high hopes, but the technologies must be packaged with deliberate, transformative policies to achieve net-zero emissions and prioritize roadway users of all types. We can apply connected and automated technologies to infrastructure, systems, and/or services that specifically support climate-friendly policies. Sometimes, innovation comes from building new services or re-purposing existing ones. Other times, innovation comes from nudging away from services.

However, even as there is a clear need for considering CCAM solutions in SUMP processes, its purpose should not be misunderstood as endorsing the disruptive technologies surrounding connected and automated technologies and their impacts, but rather empowering the local authorities to critically review the anticipated technological changes and shape the future according to their expectations and their citizens mobility needs.

Using a design thinking approach this research focused on new ways that connected and automated technologies can address climate concerns. It proposed concepts including the ideas that:

- Connect and automated technologies might enforce non-car-centric climate-friendly policies (e.g., automatic bus lane or speed limit enforcement).
- AVs could leverage connected and automated suburban freight system to provide community services during disaster.
- Fleets of electric AVs might service mobile charging stations serving networks of rural areas and assist in grid resilience and load balancing.

These concepts warrant further study going into the future. While have the potential to significantly impact cities, their infrastructure, and their residents, they can clearly be used as a tool to increase accessibility and safety alongside climate resilience. With appropriate policy guidance planners, engineers and policy makers will ensure that they are deployed in a way that maximizes their benefits for everyone.

References

1. European Environment Agency. https://www.eea.europa.eu/en/topics/in-depth/transport-and-mobility
2. Boeing, G., (Billy) Riggs, W.: Converting one-way streets to two-way streets to improve transportation network efficiency and reduce vehicle distance traveled. J Planning Educ. Res., July 2022. https://doi.org/10.1177/0739456X221106334
3. van Arem, B., et al.: Building automation into urban and metropolitan mobility planning. In: Road Vehicle Automation 6, Meyer, G., Beiker, S. (eds.) in Lecture Notes in Mobility, pp. 123–136. Springer International Publishing (2019)
4. Riggs, W., Beiker, S.A.: Business models for shared and autonomous mobility. In: Meyer, G., Beiker, S. (eds.) AVS 2019. LNM, pp. 33–48. Springer, Cham (2020). https://doi.org/10.1007/978-3-030-52840-9_4
5. Etminani-Ghasrodashti, R., Ketankumar Patel, R., Kermanshachi, S., Michael Rosenberger, J., Weinreich, D., Foss, A.: Integration of shared autonomous vehicles (SAVs) into existing transportation services: A focus group study. Transp. Res. Interdisciplinary Perspectives **12**, 100481 (2021). https://doi.org/10.1016/j.trip.2021.100481
6. Fagnant, D.J., Kockelman, K.M.: The travel and environmental implications of shared autonomous vehicles, using agent-based model scenarios. Transp. Res. Part C: Emerg. Technol. **40**, 1–13 (2014)
7. Rupprecht Consult - Forschung & Beratung GmbH (eds). 2019. Topic Guide: Road vehicle automation in sustainable urban mobility planning. https://www.eltis.org/mobility-plans/topic-guides
8. European Commission. Expectations and Concerns from a Connected and Automated Mobility. https://europa.eu/eurobarometer/surveys/detail/2231
9. NetZeroCities project. https://netzerocities.eu/the-nzc-project/
10. Elvik, R., Meyer, S. F., Hu, B., Ralbovsky, M., Vorwagner, A., Boghani, H.: Methods for forecasting the impacts of connected and automated vehicles, Deliverable D3.2 of the H2020 project LEVITATE (2020)
11. Milakis, D., van Arem, B., van Wee, B.: Policy and society related implications of automated driving: a review of literature. J. Intell. Transp. Syst. **21**(4), 324–348 (2017)
12. Rupprecht Consult - Forschung & Beratung GmbH. 2020. Enabling "Automation-Ready" Transport Planning. How to become an Automation-Ready road authority? Available at https://h2020coexist.wpenginepowered.com/wp-content/uploads/2020/05/D1.2_Automation-Ready-Framework.pdf
13. Snelder, M., Wilmink, I., van der Gun, J., Bergveld, H.J., Hoseini, P., van Arem, B.: Mobility impacts of automated driving and shared mobility – explorative model and case study of the province of North-Holland. Europ. J. Transp. Infrastructure Res. **9**(4), 291–309 (2019)
14. COWI and PTV (2019), The Oslo study – How autonomous cars may change transport in cities', Oslo (2019). https://ruter.no/globalassets/dokumenter/ruterrapporter/2019/the-oslo-study.pdf
15. ITF/OECD (2015), Urban mobility system upgrade – How shared self-driving cars could change city traffic, 15cpb_self-drivingcars.pdf (itf-oecd.org)
16. ITF/OECD (2018): The shared-use city: Managing the curb, available at shared-use-city-managing-curb_5.pdf (itf-oecd.org)
17. Riggs, W., Niel, S. Shukla, Mark, S.: The Trip Characteristics of a Pilot Autonomous Vehicle Rider Program: Late Night Service Needs and Desired Increases in Service Quality, Reliability and Safety. Rochester, NY, Aug. 01, 2022. https://doi.org/10.2139/ssrn.4195380
18. Riggs, W., Pande, A.: On-demand microtransit and paratransit service using autonomous vehicles: Gaps and opportunities in accessibility policy. Transp. Policy **127**, 171–178 (2022). https://doi.org/10.1016/j.tranpol.2022.07.024

19. Riggs, W., Pande, A.: Gaps and opportunities in accessibility policy for autonomous vehicles. Mineta Transp. Inst. (2106), Aug. 2021. Accessed: Sep. 08, 2021. https://transweb.sjsu.edu/research/2106-Accessibility-Policy-Autonomous-Vehicles
20. W. Riggs, N. Larco, G. Tierney, M. Ruhl, J. Karlin-Resnick, and C. Rodier, "Autonomous Vehicles and the Built Environment: Exploring the Impacts on Different Urban Contexts," in Road Vehicle Automation 5, G. Meyer and S. Beiker, Eds., in Lecture Notes in Mobility, pp. 221–232. Springer, Cham (2019). doi: https://doi.org/10.1007/978-3-319-94896-6_19
21. Bartholomew, K., Ewing, R.: Land use-transportation scenarios and future vehicle travel and land consumption: a meta-analysis. J. Am. Plann. Assoc. **75**(1), 13–27 (2008). https://doi.org/10.1080/01944360802508726
22. Litman, T.: Autonomous Vehicle Implementation Predictions, Victoria Transport Policy Institute, vol. 28, 2014, Accessed: Apr. 13, 2016. http://sh.st/st/787f28ed3e745c14417e4aec27303038/http://www.vtpi.org/avip.pdf
23. Crute, J., Riggs, W., Chapin, T., Stevens, L.: Planning for Autonomous Mobility. American Planning Association, Washington D.C., PAS 592 (2018). https://www.planning.org/publications/report/9157605/
24. CBS, Reading guide - Monitor of Well-being & the Sustainable Development Goals (2022). https://longreads.cbs.nl/monitor-of-well-being-and-sdgs-2022/
25. Vonk Noordegraaf, D., Wilmink, I., Bouma, G.: Indicatoren voor brede welvaart in het mobiliteitsdomein – een vertrekpunt voor discussie gebaseerd op een quickscan, TNO-rapport nr. TNO 2021 R12422, 10 december 2021 (In Dutch: Indicators of welfare beyond GDP in the mobility domain - a starting point for discussion based on a quick scan) (2021)
26. Backhaus, W.: Planning for net zero cities; presentation at ARTS 2022, 20 July 2022
27. Eltis. The SUMP Concept. https://www.eltis.org/mobility-plans/sump-concept
28. European Commission. New transport proposals target greater efficiency and more sustainable travel. https://transport.ec.europa.eu/news/efficient-and-green-mobility-2021-12-14_en
29. Eltis. SUMP topic Guide on "Decarbonisation of urban Mobility. https://www.eltis.org/in-brief/news/sump-topic-guide-decarbonisation-now-available

Check for updates

Enhancing Mobility with Automated Shuttles and Buses

Katherine F. Turnbull(✉)

Texas A&M Transportation Institute, 3135 TAMU, College Station, Bryan, TX, USA
k-turnbull@tti.tamu.edu

Abstract. This chapter presents information on demonstrations, pilots, and deployments of automated shuttles, with a focus on enhancing mobility for all users. Automated shuttles continue to be introduced in downtown areas, university campuses, business parks, entertainment complexes, and other areas. Automated buses are being tested in selected corridor applications. The chapter is based on the presentations and discussions at a breakout session at the 2022 Transportation Research Board (TRB) Automated Road Transportation Symposium (ARTS). The ARTS breakout session, Enhancing Mobility with Automated Shuttles and Buses, highlighted projects addressing the needs of all users, including individuals using wheelchairs, those with limited or no eyesight, and those with other disabilities. The information presented in this chapter will assist in evaluation of this mobility option to help inform decision making, identify research needs, and support future developments.

Keywords: autonomous shuttles · driverless shuttles · automated shuttles · automated buses · autonomous buses · driverless buses

1 Introduction

Pilots, demonstrations, and deployments of automated shuttles and buses in the United States and other countries were restored in mid-2021 after many projects were put on hold during the pandemic. New projects were also initiated. These services focus on enhancing mobility and accessibility on regular routes, providing first- and last-mile trips, and improving transportation options for individuals with disabilities. The ARTS 2022 breakout session highlighted projects in Minnesota, Florida, Texas, and Connecticut, as well as Scotland. An update was also provided on activities sponsored by the Federal Transit Administration (FTA) of the U.S. Department of Transportation (USDOT). The presentations built on previous breakout sessions in 2019, 2020, and 2021 (1, 2, 3). The presentations highlighted the participation of the numerous public agencies, private shuttle companies, and technology and community-based groups needed for successful projects. Participants in the session shared experiences with automated shuttles and buses, highlighted outreach efforts and use by disabled individuals, and described lessons learned and tips for others interested in similar applications.

G. Meyer and S. Beiker (Eds.): ARTSymposium 2022, LNMOB, pp. 72–78, 2023.
https://doi.org/10.1007/978-3-031-34757-3_7

2 Examples of Automated Shuttle and Bus Pilots, Demonstrations, and Research

2.1 Autonomous Shuttle Pilots in Minnesota

The Minnesota Department of Transportation (MnDOT) has conducted numerous research projects and pilots focusing on automated shuttles. Testing an EasyMile automated shuttle under winter weather conditions at the MnROAD research facility represented an initial project. The shuttle was also operated on the Nicollet Mall in downtown Minneapolis during Super Bowl Week in January 2018. MnDOT hosted additional events introducing the automated shuttle to diverse groups, including members of the Minnesota Chapter of the National Federation for the Blind.

The Med City Mover represents the second pilot conducted by MnDOT. It included the 12-month operation of two EasyMile EZ10 vehicles in downtown Rochester, which is the home of the Mayo Clinic. MnDOT was the project lead. The project partners included the City of Rochester, the Mayo Clinic, and Destination Medical Center. First Transit and EasyMile were the technology and transit operations partners.

The project goals focused on engaging Minnesotans about the potential benefits and opportunities of AV technology, improving the operation of AVs in winter weather conditions, and identifying changes to infrastructure needed to safely operate AVs on public roadways. Another goal was enhancing the transportation experience for Rochester residents, businesses, and visitors, and improve how people get around in the high-demand downtown area.

The two six-passenger electric EasyMile E210 shuttles operated on a 1.5-mile-long loop, connecting the Mayo Clinic Downtown Campus with restaurants, grocery stores, residential areas, apartment complexes, hotels, and parking facilities. The route included stops at two locations. The free service was provided from 9:00 a.m. to 3:00 p.m., Monday – Friday, and 9:00 a.m. to 5:00 p.m. Saturday and Sunday.

An onboard ambassador was present on the shuttle to assist passengers and to take over operation if needed. Vehicle and project accessibility features included wheelchair ramps and wheelchair tie-downs, signage in Braille, and the use audio messages, trolly bells, and video with closed captioning to communicate with disabled passengers.

Pre- and post-ride surveys conducted by MnDOT highlighted the benefits of the pilot as a way to build familiarity and acceptance of AVs among the public. In the pre-ride survey, 87 percent of the respondents reported being "very positive," "mostly positive," or "somewhat positive" about AVs in Minnesota. In the post-ride surveys, the responses in these categories increased to 97 percent. The total of "very positive" responses alone increased from 45 to 62 percent. Riders also reported less concerns about AVs in the post-ride survey.

MnDOT is sponsoring two other AV shuttle research and demonstration pilots. One pilot is in the city of White Bear Lake, a suburb to the north of the city of St. Paul. The second pilot is in the city of Grand Rapids and the surrounding area in north-central Minnesota. These two pilots are using different technologies, and operating strategies.

The Bear Tracks automated shuttle in White Bear Lake operates on a 1.5-mile route, connecting the community YMCA, affordable housing, and a center offering day programs for adults with intellectual and developmental disabilities. Project goals focus on operating an AV shuttle in a suburban setting in all-weather conditions, increasing transportation options for those facing transportation barriers, engaging the public to build awareness of AV shuttles, and supporting workforce development.

The 12-month pilot began in August 2022. Bear Tracks operates on weekdays, from 9:30 a.m. to 1:30 p.m. In addition to MnDOT, other project partners include AECOM, NEWTRAX, NAVYA, Ramsey County and the city of White Bear Lake.

The goMARTI (Minnesota Autonomous Rural Transportation Initiative) offers on-demand rides in a 17 square mile area in the Grand Rapids area. Service is provided using five May Mobility automated vehicles, including three wheelchair-accessible vehicles. Project goals focus on advancing the operation of automated vehicle technology in rural winter weather conditions, engaging and educating the local community, and providing safe, accessible mobility for residents, especially those with transportation challenges. Another project goal is to understand the potential economic development benefits the innovative pilot brings while attracting future talent and technology to the area.

Service operates Tuesday through Friday from 2:00 p.m. to 10:00 p.m., Saturdays from 10:00 a.m. to 10:00 p.m., and Sundays from 8:00 a.m. to 2:00 p.m. Riders can book trips through a smartphone app or by calling 221. Project partners include MnDOT, May Mobility, PLUM Catalyst, the city of Grand Rapids, Via, First Call 211, the Iron Range Resources and Rehabilitation Board, Itasca County, the Blandin Foundation, Mobility Mania, and Visit Grand Rapids. Other Collaborative partners include the University of Minnesota, MnCAV Ecosystem, Arrowhead Regional Development Commission, IASC NEXT Career Pathways, and Innovate 218.

2.2 I-STREET Living Lab and Autonomous Shuttles in Florida

Implementing Solutions from Transportation Research and Evaluating Emerging Technologies (I-STREET) is a collaboration among the University of Florida (UF), the Florida Department of Transportation (FDOT), and the city of Gainesville. It provides a real-world CAV testbed in Gainesville focusing on safety, mobility, data analytics, and human factors. It includes facilities and projects in the Gainesville area and across Florida.

The Gainesville autonomous shuttle pilot represents a partnership with UF, FDOT, and the city of Gainesville. Local stakeholders include community groups. Beep, Transdev, and EasyMile are the service and vehicle providers.

The autonomous shuttle route connects the UF campus and downtown Gainesville. Before and after surveys were conducted to gain insights into the perception of riders. The before survey was conducted online and in person in the summer of 2018. Service was stopped in 2020 and restarted in January 2021. The after survey was conducted online in the spring of 2021. The surveys included questions on travel behavior and technology use, autonomous shuttle comfort and safety, and demographics. Perceptions related to riding in the shuttles, including the ability of the service operator to take control of the shuttle if needed were much more positive in the after surveys once individuals rode in the vehicles.

Florida I-STREET also examined the Beep shuttle operating at Lake Nona. The service, using a NAVYA shuttle, launched in September 2019. There was a pause in service due to the pandemic in March 2020, with service relaunched in June 2020.

Traffic behavior was observed using recorded video data at a crosswalk, a signalized intersection, and an all-way stop controlled intersection.

The I-STREET and Lake Nona projects indicate that travelers became more comfortable with automated shuttles once they experience the technology. The projects also highlight the importance of partner and stakeholder involvement and interaction.

2.3 Automated Shuttles and Buses for All Users Research Project, Texas a&M Transportation Institute

A research project, Automated Shuttles and Buses for All Users, is being conducted by the Texas A&M Transportation Institute (TTI) as part of the Safety Through Disruption (SAFE-D) University Transportation Center (UTC) led by the Virginia Tech Transportation Institute. The project introduced individuals with mobility and visual impairments to AVs and a smart intersection and gained information on their complete trip. The project is identifying improvements in AVs, service operations, the street system, and the built environment to ensure that individuals with disabilities have equal and safe access to automated shuttles and buses to improve their mobility.

In cooperation with the city of Arlington, the University of Texas Arlington, Via Rideshare, and May Mobility, the project introduced individuals with mobility and visual impairments to the Rideshare, Automation, and Payment Integration Demonstration (RAPID) automated shuttles in June, 2021. The sessions included a pre-interview, a ride in a RAPID shuttle, and a post-interview. In addition, online interviews were conducted in the spring of 2022 with Texas A&M University students using mobility devices to obtain their reactions to automated shuttles and a smart intersection which alerts pedestrians of buses making left turns.

The overall initial reactions to riding in the RAPID shuttles from the individuals was very positive. Individuals noted they felt comfortable and safe riding in the vehicles. They noted the generally smooth ride, including when the vehicle was in automated operation. Some participants used assistance to enter and exit the vehicles, and they noted the importance of having an assistant for regular use. Responses from the online interviews to automated shuttles was also positive, as was the response to the smart intersection. Participants in both groups noted they would use automated shuttle services on a regular basis if available.

Participants in both groups provided insights into their complete trips and the built environment. They noted the importance of handicapped vehicle parking spots, as well as curb cuts, ramps, and accessible sidewalks. The need to lower traffic speeds in many areas was noted. Ensuring the correct locations for pickups and drop-offs and the need for on-vehicle attendants to provide assistance was stressed as important. The project is developing guidelines for use in ensuring that individuals with disabilities have equal access to automated shuttles.

2.4 CT*Fastrak* CAV Bus Project

The CT*Fastrak* is a 9.4-mile-long bus rapid transit (BRT) corridor between Hartford and New Britain owned and maintained by the Connecticut Department of Transportation (CTDOT). The bus-only roadway includes 11 stations and five at-grade intersections.

There is also a multi-use trail along a five-mile segment, providing seasonal access at 12 locations. Frequent bus service is operated by CTtransit. The CT*Fastrak* CAV Bus project is testing the use of three 40-foot automated battery elective buses on the bus-only roadway.

The project is funded through an FTA Integrated Mobility Innovation (IMI) grant, and an FTA Low-No grant, and other federal and state programs. CTtransit is the project lead. New Flyer of America is the electric bus manufacturer and cte is the project manager and technical consultant. Robotic Research is the technology supplier/integrator and the University of Connecticut is providing data collection and analysis assistance. The USDOT Volpe Center is also providing ongoing technical support.

The automated buses will be operated in revenue service. A safety operator, with the ability to take over operation of the bus if needed, will be in the driver's seat at all times. Automated steering, braking, and lane keeping will be tested, along with pedestrian and object detection. Traveling up to 40 miles per hour (mph), the three buses will be tested during the day and night in all weather conditions. Precision docking of buses at stations to allow for easier wheelchair access and bus platooning are other project elements. Improvements are also being made at some intersections to address line-of-site issues. Improvements include installing radar and cameras, new traffic signal controllers, and new communications technology.

Project goals include improving Americans with Disabilities Act (ADA) accessibility at platforms through precision docking, increasing vehicle efficiency and capacity on the guideway through bus platooning, and reducing the number of incidents resulting in injury or vehicle damage at two intersections due to traffic not stopping at red lights. Testing of the CT*Fastrak* CAV buses is scheduled to begin in 2023. The demonstration will occur during 2023 and 2024.

2.5 CAVForth, Edinburgh Scotland

This project is operating a fleet of five automated buses in high-capacity service on a 14-mile route across the Forth Road Bridge that links the Ferrytoll park-and-ride facility with the Transport Hub at Edinburgh Park Station. CAVForth project partners include Fusion Processing, Stagecoach, Bristol Robotics Laboratory, and Edinburgh Napier University. While the project was put on hold during the pandemic, service was launched in 2022.

The societal research plan for the project included stakeholder consultation involving an initial online survey and online workshop and an online bus passenger survey. A total of 1,054 responses were received to the bus passenger survey. The survey results seem to indicate that there may be a balance between individuals who may use the bus more because of technology and those who may ride the bus less due to technology. Older and female passengers appear to be less convinced about the benefits of automated buses. It also appears that the comfort of the ride continues to be a key motivation for use.

2.6 Federal Transit Administration Transit Bus Automation Research

The FTA Strategic Transit Automation Research (STAR) Plan (4) was published in 2018. The plan focuses on enabling research, integrated demonstrations, and strategic partnerships. STAR activities have improved transit agency understanding of technology and its maturity, opportunities, and challenges. A strong network of agency and industry representatives has also been established. In 2022, the FTA conducted a request for information to help inform the development of the STAR Plan 2.0.

Numerous programs within the FTA and the USDOT focus on ensuring accessible automated shuttles and buses. Numerous FTA-managed transit bus automation demonstrations and pilots are underway throughout the country using a variety of federal, state, and local funding sources. Examples of these programs include the FTA Accelerating Innovative Mobility and Integrated Mobility Innovation grants, and the USDOT Inclusive Design Challenge, ITS4US, and the ASPIRE University Transportation Center. The National Highway Transportation Safety Administration sponsored Automated Wheelchair Tiedown and Occupant Restraint System project provides an additional example. Research topics have focused on boarding and aligning technologies, securement technologies, and wayfinding and communication technologies. More information on these activities and other related programs is available on the FTA website (5).

3 Private Sector Activities with Automated Shuttles and Buses

A variety of business models are being used to plan and operate automated shuttles and buses in the various pilots and demonstrations. The private sector is actively involved in all projects. Automated vehicle companies, transit service operators, public engagement consultants, land development companies, and businesses are examples of the diverse private sector interest in enhancing mobility through automated shuttles and buses.

Electric automated shuttles developed by EasyMile, Local Motors, NAVA, Drive.ai, Lexus hybrid sedans, and Polaris GEM vans have been used in various pilots and demonstrations. First Transit, TransDev, and Beep provide examples of transit service companies operating the services. Property development companies, major employers, and businesses have also partnered on demonstration projects. These groups bring technical and operating expertise, resources, and business strengths to the various groups.

4 Additional Research

Topics for further research projects, pilots, and evaluation have been highlighted through projects, studies, and discussions at the AVS breakout sessions. The need for further research focusing on the use of automated shuttles and buses by disabled individuals was noted. Outreach to the disabled community was highlighted, along with conducting more pilots and demonstrations addressing the needs of disabled individuals. Continuing to examine boarding and alighting technologies, securement technologies, way finding and communication technologies, on-board assistance needs, and improvements to the built environment represents a few examples of areas for further research. Continuing to share experiences with pilots, demonstrations, and deployments was also noted as important.

References

1. Turnbull, K., Jones, C., Elefteriadou, L.: Catching up with low-speed autonomous shuttles. In: Meyer, G., Beiker, S. (eds.) AVS 2019. LNM, pp. 63–70. Springer, Cham (2020). https://doi.org/10.1007/978-3-030-52840-9_6
2. Turnbull, K., Jones, C., Elefteriadou, L.: Autonomous shuttles and buses: from demonstrations to deployment. In: Meyer, G., Beiker, S. (eds.) AVS 2020 2020. LNM, pp. 73–80. Springer, Cham (2022). https://doi.org/10.1007/978-3-030-80063-5_7
3. Turnbull, K.F.: Automated shuttles and buses for all users. In: Meyer, G., Beiker, S. (eds.) Road Vehicle Automation 9. ARTSymposium 2021. Lecture Notes in Mobility. Springer, Cham (2023). Doi:https://doi.org/10.1007/978-3-031-11112-9_9
4. Federal Transit Administration. Strategic Transit Automation Research Plan. U.S. Department of Transportation (2018)
5. https://www.transit.dot.gov/automation-research

Innovation Strategies and Research Trends for Connected, Cooperative and Automated Mobility in Europe

Carolin Zachäus[1(✉)] and Stephane Dreher[2]

[1] Department European and International Business Development, VDI/VDE Innovation + Technik GmbH, Steinplatz 1, 10623 Berlin, Germany
carolin.zachaeus@vdivde-it.de

[2] Department Innovation & Deployment, ERTICO – ITS Europe, Avenue Louise 326, 1060 Brussels, Belgium
s.dreher@mail.ertico.com

Abstract. Exploring the forefront of innovation strategies and research trends, the future of connected, cooperative, and automated mobility (CCAM) in Europe promises to revolutionize the transportation landscape, paving the way for sustainable, efficient, and inclusive transportation systems [1]. Therefore, this chapter discusses the challenges and solutions for the deployment of CCAM technologies in Europe related in particular to coordination and cross sector stakeholder engagement. Despite the technical advancements and numerous Research and Innovation (R&I) and testing activities, multiple barriers remain, such as limited demand, lack of technical maturity, scattered research efforts, and inadequate demonstration and scale-up. The CCAM Partnership plays a crucial role in overcoming these obstacles by creating a unified, long-term R&I agenda. Under the umbrella of the CCAM Partnership 18 projects have been funded so far. The project objectives and thematic assignments corresponding to R&I priorities identified by the Partnership, are described in this paper. Furthermore, the role of the FAME project supporting the European CCAM ecosystem by fostering knowledge exchange, best practices, and international collaboration is highlighted. Finally, the CCAM ecosystem with respect to the R&I projects will be discussed.

Keywords: Connected and Automated Mobility · Autonomous Driving · Electronic Components and Systems · Innovation Policy · Research and Innovation Activities · CCAM Partnership

1 Introduction

Mobility and transport are vital to society, enabling economic and social life, yet they also impose significant costs, including pollution, accidents, and greenhouse gas emissions. The transport sector must significantly reduce its emissions to achieve EU climate goals, and the European Green Deal's success depends on sustainable transport [2]. As society moves towards sustainable and efficient mobility solutions, CCAM has emerged as a

G. Meyer and S. Beiker (Eds.): ARTSymposium 2022, LNMOB, pp. 79–90, 2023.
https://doi.org/10.1007/978-3-031-34757-3_8

crucial area of focus. CCAM solutions aim to provide user-centred, inclusive mobility while increasing safety, reducing congestion and emissions, and contributing to decarbonization. By seamlessly integrating novel mobility services with existing ones, they have the potential to enhance transport sustainability, safety and inclusiveness directly contribute to various policy goals, such as the United Nations Sustainable Development Goals, Vision Zero, the European Green Deal, Europe fit for the Digital Age, and the Smart and Sustainable Mobility Strategy. However, the problem drivers and technical maturity of CCAM solutions must be further advanced and proven.

Recognizing the necessity for collaboration in the development of technological solutions, the Horizon Europe Framework strategy emphasizes the importance of European Partnerships [3]. European Partnerships in the Horizon Europe framework are strategic, collaborative initiatives designed to address global challenges, enhance European competitiveness, and foster innovation. These partnerships are collaborative initiatives between the EU Commission and private and/or public partners that support research and innovation programs and significantly contribute to achieving EU political priorities. By pooling resources, expertise, and knowledge, European Partnerships drive impactful research and innovation, accelerating progress towards a more sustainable, inclusive, and prosperous future for Europe.

Besides emphasizing the importance of European Partnership, the Horizon Europe Framework strategy aims to improve the coherence of Partnerships, make them more open and transparent, and promote competitiveness and innovation, while fostering pre-competitive collaboration of stakeholders from the entire value chain, ensuring a holistic approach to address complex challenges, e.g. posed by CCAM.

This paper provides an overview of the topics and projects addressed in the first phase of the CCAM Partnership, focusing on the essential stakeholder interaction required for successful implementation. By examining the collaborative efforts and outlining the key areas of focus, we aim to provide insights into the current state of CCAM Research and Innovation strategies in Europe and highlight the crucial role that partnerships as well as Coordination and Support Actions (CSA) play in its development and deployment.

2 Research and Innovation Strategies in Europe

As society moves towards sustainable and efficient mobility solutions, CCAM has emerged as a crucial area of focus. CCAM solutions aim to provide user-centred, inclusive mobility while increasing safety, reducing congestion and emissions, and contributing to decarbonization.

Multiple problem drivers are remaining to achieve the successful deployment of these technologies, such as limited demand, underdeveloped solutions, scattered research and innovation efforts, and inadequate demonstration and scale-up. To tackle these challenges, a transformation in the mobility innovation process is crucial, with a focus on user engagement, strategic timing, and expanded outreach. The CCAM Partnership [4] plays a vital role in overcoming these obstacles, facilitating the swift adoption of CCAM technologies and services across Europe by creating a unified, long-term research and innovation agenda that incorporates various stakeholders in a coordinated, holistic approach. This partnership ultimately supports Europe's position as a leader in safe and sustainable road transport through the advancement of automation. Launched in 2021, the European CCAM Partnership has the main objectives to better align EU R&I efforts in the field of CCAM and develop and implement a coherent long-term agenda to coordinate investments in R&I and pre-deployment. The second phase of the partnerships focusses on implementing large-scale demonstrations of inclusive and user-oriented CCAM solutions for people and goods across Europe by 2030.

Partnerships should however not work in silos, as several challenges they address are connected and in particular, for transport the wider impacts or all related technologies should be looked at. Emphasizing a collaborative approach between different partnerships and missions, the CCAM Partnership, the 2ZERO Partnership [5], and the Cities Mission [6] have jointly launched a common research call to broaden the EU perspective and enhance cooperation across various mobility domains and urban-related topics.

The calls primary objective is to develop cutting-edge mobility solutions catering to the needs of both people and cities while aiming for climate neutrality by 2030. Through collaboration with local authorities, citizens, and stakeholders, the projects funded in the frame of this call will create transferable solutions that merge electrification, automation, and connectivity in passenger and freight transport. These solutions should be economically viable, modular, adaptable, and applicable across cities committed to achieving climate neutrality.

Moreover, the project will bolster capacity among local authorities, users, and mobility systems providers, accelerating the adoption of shared, smart, and zero-emission solutions. It will also aid in formulating implementation plans for local and regional transport authorities to replicate innovative smart mobility solutions and associated infrastructure in cities beyond the project's reach.

Another development strongly discussed in the European CCAM ecosystem is the promotion of a software defined vehicle (SDV) platform. The increasing importance of software and hardware in vehicles, as they become more electric, autonomous, connected, and service-oriented, has led to a rapid rise in software complexity. This development calls for enhanced standardization and reuse through a software platform, which encompasses the operating system and middleware layer. European companies, facing intense global competition and talent shortages in automotive software, often encounter fragmented efforts.

A pre-competitive European collaboration could expedite cooperation and result in an open SDV platform, initiated by the Electronic Components and Systems Partnership (KDT) and spearheaded by leading EU stakeholders and concentrating on non-differentiating aspects [7]. This approach would help conserve resources and focus investments on competitive solutions. The platform aims to establish common architecture building blocks (modules), supported by development tools for prototyping and testing, centering on the hardware abstraction layer within a comprehensive SDV architecture.

3 Research Landscape in Europe (Stakeholder and/or Projects)

The European Union has a long legacy of funding research and innovation activities in CCAM. Intense funding with the launch in 2016 of a specific call on "Automated Road Transport" in the Horizon 2020 Programme provided over €300 million in funding up to 2020 in four research fields, namely: Networking, Coordination & Support; Infrastructure, Connectivity and Cooperative Systems; Driver Assistance Systems and Partial Automation; and Highly Automated Road Transport. EU research funding has continued in the Horizon Europe programme for 2021–2027 under the umbrella of the CCAM Partnership. So far, 18 projects have been funded in the frame of the Horizon Europe Programme in CCAM. Each project is assigned to one of the seven clusters, which structure the activities of the CCAM Partnership (Fig. 1).

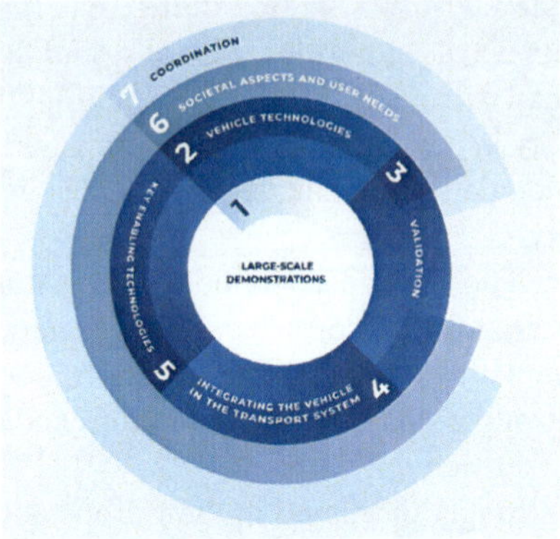

Fig. 1. Cluster structure of the CCAM Partnership, organizing the necessary R&I actions to support large-scale demonstration essential to advance towards deployment readiness.

The interlinked CCAM cluster structure illustrates the connections between specific R&I actions and the Partnership's objectives to advance CCAM solutions and prepare them for large-scale deployment.

The successful deployment of CCAM relies among other things highly on its societal benefits and user adoption. To achieve this, development, deployment, and regulation must be based on understanding specific needs, impacts, and costs. Cluster 6 addresses user needs and societal aspects in various ways, such as focusing user-centric technologies, guiding transport system integration, addressing user needs in key enabling technologies, and offering feedback on societal aspects in Living Labs. Socio-economic and environmental impacts will be assessed to understand CCAM's contributions to safety, accessibility, equity, and environmental goals. Tools will be provided for user-centered solutions that effectively contribute to societal targets and regional CCAM uptake. Concrete actions will be performed in the following projects, which have started in September 2022.

MOVE2CCAM - MethOds and tools for comprehensive impact Assessment of the CCAM solutions for passengers and goods [8] explores the impact of CCAM passenger and freight solutions by defining use cases, business models, and KPIs through co-creation with a multi-system network of actors. It develops a system dynamics-based impact assessment tool to evaluate the effects of CCAM interventions on various aspects, considering European region specifics and different actors' needs. The project delivers impact evaluation frameworks, KPIs, policy recommendations, and recommendations for Sustainable Urban Mobility Plans (SUMP).

SINFONICA - Social INnovation to FOster iNclusIve Cooperative, connected and Automated mobility [9] aims to develop strategies, methods, and tools to engage CCAM users, providers, and stakeholders in understanding their needs and concerns related to CCAM. It co-creates decision support tools for designers and decision-makers to enhance seamless and sustainable CCAM deployment, ensuring inclusivity and equity for all citizens.

Besides the understanding of user needs and societal aspects, the advancement of vehicle technologies for sensing and safety systems (cluster 2), alongside with key enabling technologies (cluster 5) are crucial to enhance CCAM solutions.

Cluster 2 "Vehicle technologies" aims to develop safe, efficient, and effective solutions for highly automated vehicles in Europe's future mobility and transport system. Cluster 2 focuses on environmental perception and safe decision making to ensure safe interactions with other road users, provide protection in emergencies, and maintain occupants' comfort and well-being. This requires multiple sensing devices and systems interacting within a "Sense-Think-Act" process for effective decision-making.

AWARE2ALL - Safety systems and human-machine interfaces oriented to diverse population towards future scenarios with increasing share of highly automated vehicles [10] addresses new safety challenges posed by highly automated vehicles in mixed traffic by developing innovative passive and active safety and Human Machine Interface (HMI) systems. It proposes a universal safety framework for HMI, building on results from previous projects and focusing on the variety of the population.

EVENTS - ReliablE in-Vehicle pErception and decisioN-making in complex environmenTal conditionS [11] aims to create a robust and self-resilient perception and decision-making system for autonomous vehicles (Avs) to handle unexpected situations. The project focuses on three use cases: Interaction with Vulnerable Road Users (VRU), Non-Standard and Unstructured Road Conditions, and Low Visibility and Adverse Weather Conditions.

ROADVIEW - Robust Automated Driving in Extreme Weather [12] develops an in-vehicle system for advanced environment and traffic recognition, prediction, and decision-making under various conditions, including harsh weather. The project integrates a cost-effective multisensory setup, sensor noise modeling, collaborative perception, and testing through simulation-assisted methods.

Cluster 5 "Key enabling technologies" supports the whole mobility system, focusing on AI, Big Data, and cybersecurity for vehicle technologies, integration, and validation, extending their application beyond individual vehicles. It fosters cooperation among stakeholders from various technology areas and industries, aiming for safe and secure operation of vehicles and mobility systems.

AI4CCAM - Trustworthy AI for CCAM [13] will develop an open environment for integrating trustworthy AI models for VRU behavior anticipation in urban traffic conditions, focusing on road safety and user acceptance.

AITHENA - AI-based CCAM: Trustworthy, Explainable, and Accountable [14] will research explainable AI in CCAM development and testing frameworks, focusing on data, models, and testing.

CONNECT - Continuous and Efficient Cooperative Trust Management for Resilient CCAM [15] addresses security and safety convergence in CCAM by assessing dynamic trust relationships and defining a trust reasoning framework, enabling cyber-secure data sharing and trustworthy outsourcing of tasks.

SELFY – SELF assessment, protection & healing tools for a trustworthY and resilient CCAM [16] aims to increase CCAM ecosystem safety, security, robustness, and resilience by developing a toolbox of collaborative tools focusing on situational awareness, data sharing, resilience, and trust. These tools will operate individually or cooperatively to manage protection, response, and recovery decisions locally or globally in response to cyber threats or hazards.

Integrating the overall transport system ensures safe human-machine interaction and supports traffic, fleet management, and physical and digital infrastructure requirements. In this regard, Cluster 4 "Integrating the vehicle in the transport system" advances physical and digital infrastructure, connectivity, and cooperation to enhance fleet and traffic management systems for CCAM vehicles. Hereby, Cluster 4 focuses on providing digital information, developing connectivity and communication solutions, delivering cybersecurity and data sharing approaches, while addressing user needs and societal expectations. Those developments are aimed to enhance safety and efficiency as well as interoperability, ensuring seamless mobility across various operators and service providers.

AUGMENTED CCAM - Augmenting and Evaluating the Physical and Digital Infrastructure for CCAM deployment [17] aims to advance the readiness of physical, digital, and communication infrastructure for large-scale deployment of CCAM solutions. It will develop and evaluate physical, digital and communication infrastructure (PDI) supported solutions in seven test sites across three European countries. AI, Big Data, and crowdsourced High Definition (HD) maps will enhance situational awareness, prediction, and actuation.

CONDUCTOR - Fleet and traffic management systems for conducting future cooperative mobility [18] focuses on designing and demonstrating advanced traffic and fleet management for efficient and globally optimal transport. It will build upon existing CCAM solutions, using dynamic balancing and priority-based management. The project will lead to reduced urban traffic, congestion, pollution, and improved quality of life.

IN2CCAM - Enhancing Integration and Interoperability of CCAM eco-system [19] aims to develop, implement, and demonstrate innovative CCAM services, providing benefits such as safety, environmental impact reduction, and inclusiveness. Physical, digital, and operational infrastructures will be enhanced to improve CCAM services and traffic efficiency. The project will be implemented in four lead Living Labs across Europe.

PoDIUM - PDI connectivity and cooperation enablers building trust and sustainability for CCAM [20] identifies and assesses connectivity and cooperation enablers for higher levels of automation, using facilities from three Living Labs in Germany, Italy, and Spain. A multi-connectivity approach ensures the reliability, availability, and redundancy of the PDI system. The project focuses on integration, advanced environmental modeling, digital twins, and VRUs in the overall PDI.

Successful implementation of CCAM depends on societal acceptance and adoption, with safety assurance as a crucial factor for trust. Cluster 3 "Validation" provides procedures, methodologies, and tools for validating, verifying, and rating CCAM systems in terms of technology and human factors, establishing an EU-wide database of relevant scenarios for validation. These actions will help reduce the number of test kilometers needed for safety validation, ensure functional safety, and develop a harmonized simulation environment for virtual testing of CCAM functions and systems.

I4Driving - Integrated 4D driver modelling under uncertainty [21] aims to create an industry-standard methodology to establish a human road safety baseline for virtual assessment of CCAM systems. The project focuses on a multi-level, modular simulation library for human driving behavior and a cross-disciplinary methodology to account for uncertainties in human behaviors and use case circumstances.

SUNRISE - Safety assUraNce fRamework for connected, automated mobIlity SystEms [22] will develop and demonstrate a safety assurance framework for CCAM systems by addressing the needs of diverse use cases, defining a scenario-based database framework, generating CCAM test scenarios, preparing tools for comprehensive testing, and integrating functional safety and cybersecurity. The project will create building blocks for the framework, including harmonized safety assessment methodologies, a federated European scenario database framework, and a commonly agreed simulation framework.

With respect to demonstrating the maturity of the developed CCAM solutions, Cluster 1 "Large-scale Demonstration" focuses on implementing results from other clusters into pilots, FOTs, and Living Labs to support deployment readiness and final impact assessment. It builds on technologies and methods, integrates concepts in real-life conditions, aligns with societal needs and, aims for cross-sector collaboration, and provides feedback and lessons learned to the CCAM community.

MODI - A leap towards SAE L4 automated driving features [23] aims to identify and resolve barriers for SAE level 4 (L4) CCAM vehicles on the corridor from Rotterdam to Oslo, demonstrating solutions for logistic chains. The project emphasizes coordination to integrate CCAM into existing logistics operations and smart traffic management and creates detailed business models to demonstrate CCAM's profitability. It comprises five use cases, focusing on regulatory barriers and infrastructure on public roads.

ULTIMO - Advancing Sustainable User-centric Mobility with Automated Vehicles [24] aims to create an economically feasible and sustainable integration of Avs for Mobility as a Service (MaaS) public transportation and Logistics as a Service (LaaS) urban goods transportation. With a user-centric approach, it targets the deployment of 15 or more multi-vendor SAE L4 Avs per site in three European locations, operating without safety drivers. The project focuses on long-term sustainable impact on automated transportation, ensuring interoperability between stakeholders, and building on previous AV-demonstrator projects to maximize technical and societal impacts.

Finally, cluster 7 "Coordination" manages all CCAM stakeholders and aligns R&I activities, facilitating knowledge exchange to address the fragmentation of efforts and lack of a coherent, long-term vision and strategy. It focuses on developing harmonized approaches, common methodologies, and tools to facilitate cross-sector collaboration as well as exchange of best practices and lessons learned.

FAME - Framework for coordination of Automated Mobility in Europe [25] has been established to support the CCAM Partnership and implement the activities of cluster 7, supporting collaboration among CCAM stakeholders for large-scale demonstrations and future scale-up of complete CCAM solutions. The project supports the European Commission and the CCAM Partnership's commitment to a long-term coordination framework for R&I and large-scale testing and evaluation activities in Europe. FAME will establish a stakeholder-validated European framework for testing on public roads, including a CCAM test data space, a common evaluation methodology, and knowledge exchange mechanisms. This framework will improve cooperation, consensus building, and data sharing, enabling comparability and complementarity of results across all testing and demonstration activities in Europe.

FAME interacts with a large stakeholder community, which extends beyond the CCAM Partnership, building on the wide network federated in the previous Coordination and Support Actions CARTRE [21] and ARCADE [22]. The project engages in particular with the international community through the Trilateral EU-US-Japan Working Group on Automation in Road Transportation and with EU funded and national R&I projects and initiatives to foster the exchange of knowledge, best practices and lessons learned and build consensus on future R&I needs. The EU wide Knowledge Base and networking tools from FAME are strongly supporting the CCAM Partnership in updating the Strategic Research & Innovation Agenda (SRIA) [26], in developing the future CCAM related Horizon Europe Work Programmes. Further support is done by monitoring the progress towards its objectives and KPIs. In addition to the support of the international Trilateral exchanges, networking tools maintained by FAME include the biennial EUCAD conferences and symposia, co-organized with the European Commission and the Partnership, as well as R&I projects and stakeholder concertation workshops, discussing specific challenges related to R&I and harmonization challenges and needs among experts.

The close cooperation with member States and cities is fundamental to gather information about national, regional or local initiatives in Europe. Thanks to the fruitful collaboration with the CCAM Platform between 2019 and 2021 in the frame of ARCADE, today more than half of the about 350 projects listed in the Knowledge Base are from Member States. In FAME, a collaboration has been initiated with the CCAM Partnership Member States Representatives Group (SRG). It is part of the current objectives of Cluster 7 to support the work of this group to exchange best practices and develop harmonized approaches enabling cross border testing. FAME is currently working with the SRG to collect requirements for testing activities (incl. Small-scale pilots, large-scale demonstration sites, and living labs) in all member states. This will form a basis to investigate the different approaches, commonalities and peculiarities across countries and prepare recommendations for mutual recognition, which will be part of the European Framework for testing on public roads.

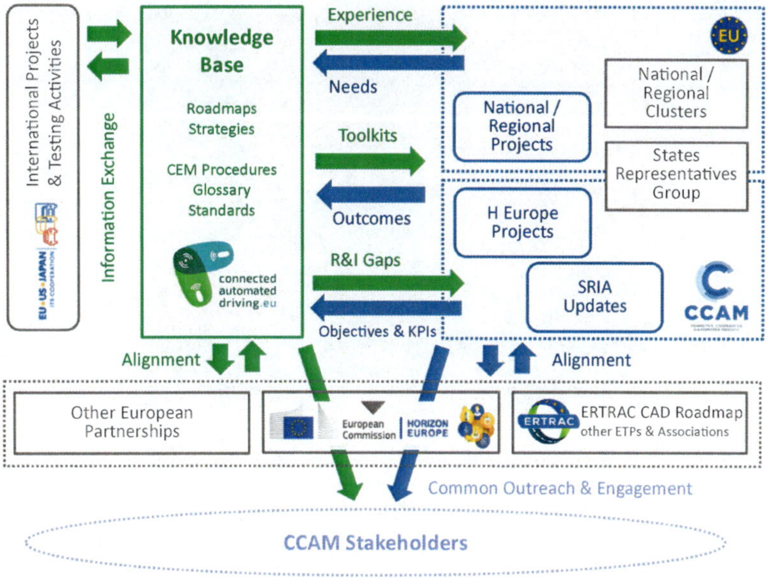

Fig. 2. FAME interaction with stakeholder ecosystem

The relationship FAME maintains with the stakeholder community is based on a win-win principle (see Fig. 2). On one side, FAME relies heavily on contributions from CCAM R&I projects, Member States, CCAM Partnership partners and the international network to carry out the tasks of collecting knowledge on ongoing activities, roadmaps, standards and common methodologies included in the Knowledge Base. Feedback from experts is also sought for the development of the Common Evaluation Methodology and related taxonomy. On the other hand. R&I project representatives and CCAM stakeholders can benefit from the experience of other projects when planning and setting up new research testing and demonstration initiatives. This bi-directional exchange will facilitate alignment and contribute in reducing overlaps or redundancies of R&I activities in Europe for a more optimized funding and coordination.

The CCAM stakeholder landscape is continuously expanding and encompasses a wide range of contributors from various sectors (Fig. 3). Key players include academia and research institutions, private industry and infrastructure providers, as well as national and regional transport authorities. The research sector, comprising universities and private research institutes, is the main driving force behind CCAM R&I projects in the framework of the CCAM Partnership under Horizon Europe. The automotive supply chain also plays a significant role, followed by physical and digital infrastructure providers. Regional and national transport authorities and stakeholder representation associations on European level contribute to the development of CCAM as well. However, public transport operators and the freight and logistics sector as well as civil society representative currently have a more limited role within the CCAM ecosystem.

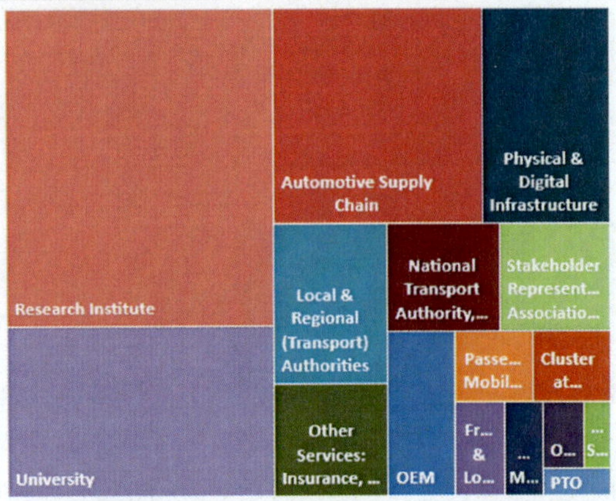

Fig. 3. Proportion of participation of different stakeholder types in CCAM project funded under the Horizon Europe framework, showing strong participation of research institutes (rose), universities (light purple), automotive suppliers (light red), physical and digital infrastructure (dark turquoise), local and regional (transport) authorities (light blue). Stakeholder representation associations on European level (light green), national transport authorities (dark red), other services (dark green) and OEMs (blue) are less represented in European R&I projects. Passenger mobility providers (orange), clusters at national level (brown), freight and logistics services (purple), national ministries (navy), civil society representation (green) and public transport operators (lavender).

Analyzing the research landscape via a network graph (Fig. 4), we can identify distinct connections within individual R&I projects, as well as key stakeholders who act as links between various R&I projects. The bigger the node, the higher the influence of the stakeholder within the research landscape. These so to say connectors are primarily research institutes in pink. European associations (grey), automotive suppliers (green), service providers (orange), and universities (black) also play significant roles. Notably, local and regional (transport) authorities, along with infrastructure providers, are not the primary connecting players, suggesting that the focus remains on developing solutions rather than implementing them.

The FAME project (pink) is strategically positioned at the center of the stakeholder network, incorporating a diverse range of stakeholder types that facilitate connections with other projects. Similarly, projects related to key enabling technologies (orange) and vehicle technologies (turquoise) are also placed closer to the center, signifying their central role in the network. Node with high centrality have a large influence on e.g. knowledge and information transfer within the network, under the assumption that the knowledge transfer follows the shortest paths. In contrast, projects associated with large-scale demonstrations (green), user-needs and societal aspects (red), and integration into the transport system (navy) are less centralized and have fewer connections. This indicates that these projects introduce new players, such as national and local authorities and societal representatives, into the network. However, it is worth noting that one project focusing on user needs and societal aspects is entirely disconnected from the rest, highlighting potential gaps in collaboration within the CCAM ecosystem.

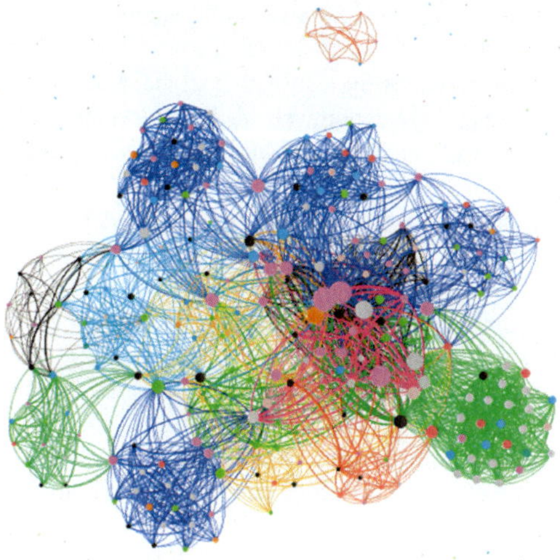

Fig. 4. Network graph, showing links between different stakeholders made through the participation in European CCAM projects. Links are colored with respect to the corresponding CCAM cluster. Knots are categorized with respect to the different stakeholder types. A few very prominent and central stakeholder act as interconnects.

4 Conclusions and Outlook

The successful deployment of CCAM technologies in Europe requires a coordinated and collaborative approach among various stakeholders, including academia, research institutions, private industry, and transport authorities. The CCAM Partnership, along with other partnerships and projects such as FAME, is essential in overcoming the challenges and facilitating the adoption of CCAM solutions.

Further alignment with other partnerships such as 2ZERO and the Cities Mission will enhance cooperation across different mobility domains and urban-related topics. Inclusion of additional stakeholders, apart from the usual suspects, can foster broader perspectives and innovative solutions. While many efforts have been made to align CCAM initiatives within Europe, there is still a need for increased focus and coordination. By fostering collaboration, standardization, and innovation across the European CCAM ecosystem, the CCAM Partnership aims to support Europe's position as a leader in safe and sustainable road transport through the advancement of automation.

Acknowledgements. The authors would also like to express their gratitude to the European Commission for funding this work through the Research and Innovation Action FAME, as well as thank the CCAM Partnership for close collaboration.

References

1. Meyer, G., et al.: Innovation strategies and funding policies for automated and electric road mobility. In: Meyer, G., Beiker, S., (Eds.), Road Vehicle Automation 9. Springer, Cham (2022). https://doi.org/10.1007/978-3-031-11112-9_5

2. European Commission, Sustainable and Smart Mobility Strategy – Putting European transport on track for the future (2020)
3. European Commission, Horizon Europe – Investing to shape our future. (2021)
4. European CCAM Partnership Association website. https://ccam.eu/. Accessed 2 Apr 2023
5. European 2ZERO Partnership Association website. https://www.2zeroemission.eu/. Accessed 2 Apr 2023
6. Mission Cities website. https://netzerocities.eu/. Accessed 2 Apr 2023
7. European KDT Joint Undertaking website. https://www.kdt-ju.europa.eu/. Accessed 2 Apr 2023
8. MOVE2CCAM website. https://move2ccam.eu/. Accessed 2 Apr 2023
9. SINFONICA website. https://sinfonica.eu/. Accessed 2 Apr 2023
10. AWARE2ALL information on CORDIS website https://cordis.europa.eu/project/id/101076868. Accessed 2 Apr 2023
11. EVENTS website. https://www.events-project.eu/. Accessed 2 Apr 2023
12. ROADVIEW website. https://roadview-project.eu/. Accessed 2 Apr 2023
13. AI4CCAM website. https://www.ai4ccam.eu/about-us/. Accessed 2 Apr 2023
14. AIthena information on CORDIS website, https://cordis.europa.eu/project/id/101076754/. Accessed 2 Apr 2023
15. CONNECT project website, https://horizon-connect.eu/. Accessed 2 Apr 2023
16. SELFY Project website, https://selfy-project.eu/. Accessed 2 Apr 2023
17. AUGMENTED CCAM Project website, https://augmentedccam.com/. Accessed 2 Apr 2023
18. CONDUCTOR Project website, https://conductor-project.eu/. Accessed 2 Apr 2023
19. IN2CCAM Project information on CORDIS website, https://cordis.europa.eu/project/id/101076791/. Accessed 2 Apr 2023
20. PODIUM Project website, https://podium-project.eu/. Accessed 2 Apr 2023
21. i4Driving project website, https://i4driving.eu/. Accessed 2 Apr 2023
22. SUNRISE project website, https://ccam-sunrise-project.eu/. Accessed 2 Apr 2023
23. MODI project information on CORDIS website, https://cordis.europa.eu/project/id/101076810/it. Accessed 2 Apr 2023
24. ULTIMO project website, https://ultimo-he.eu/. Accessed 2 Apr 2023
25. FAME Project (2023). https://www.connectedautomateddriving.eu/about/fame/. Accessed 2 Apr 2023
26. Strategic Research and Innovation Agenda 2021–2027: European leadership in safe and sustainable road transport through automation. CCAM Partnership, Brussels 2022
27. CARTRE Project on European Commission TRIMIS website, https://trimis.ec.europa.eu/project/coordination-automated-road-transport-deployment-europe. Accessed 2 Apr 2023
28. ARCADE (ConnectedAutomatedDriving) Project website https://www.connectedautomateddriving.eu/about/arcade. Accessed 2 Apr 2023

Part II: Business Models and Operations

The Trip Characteristics of a Pilot Autonomous Vehicle Rider Program: Revealing Late Night Service Needs and Desired Increases in Service Quality, Reliability and Safety

William Riggs(✉), Niel Schrage, Shivani Shukla, and Shannon Mark

University of San Francisco, 2130 Fulton Street, San Francisco, USA
{wriggs,nschrage,sgshukla,sjmark}@usfca.edu

Abstract. A substantial and growing body of literature has provided educated guesses and transportation demand modeling about how riders might behave in autonomous vehicles (AVs). No studies to-date have explored how riders behave when given access to rides in these new modes of transportation, and how AVs can help address lingering transportation challenges in the city, such as transit deserts, congestion, and increased sustainable modes of transport. This paper evaluates a first-of-its-kind program, offering passengers autonomous rides in Cruise vehicles between the hours of 11:00pm-5:00am when transit services are less prevalent. Results indicate that more than 76% of reported travel by AV riders was mode substitution, largely diverting from rideshare and transit. Over 55% of trips replaced rideshare travel—most of these trips were for social/recreational and shopping/errands. These results suggest that most AV trips may not create induced or latent demand but rather provide an opportunity to address network gaps and last mile connectivity. The results hold additional promise as the importance and popularity of new shared vehicle solutions emerge in the marketplace.

Keywords: Autonomous Vehicles · Pilot · Travel Behavior · Rider Preferences Business Models · Rideshare · Latent Demand

1 Introduction

In recent years many studies have hypothesized and developed scenarios for how individuals might use autonomous vehicles [1, 2]. Many of these studies predict trends using survey data or proxy data from rideshare experiments—for example, work from UC Davis using a chauffeur given to various families over a weeklong period to understand how they would use the vehicle [3] or hypothetical surveys that ask people if they would like to have a personally-owned autonomous vehicle for errands and explored attitudinal constructs [4].

While these predictive studies offer a lens into human perceptions, they are not reflective of true travel behavior experiments regarding AV usage, nor the anticipated roll-out and business model planning of AV fleet operators. Recent work predicts AV technology

G. Meyer and S. Beiker (Eds.): ARTSymposium 2022, LNMOB, pp. 93–107, 2023.
https://doi.org/10.1007/978-3-031-34757-3_9

will be primarily rolled out in shared and fleet-based deployments in urban centers [5], and has the potential to create new innovations in microtransit and paratransit [6–8]. However, to date, no actual AV-rider experiments have been conducted at scale. Research has explored how rideshare provides additional services to historically marginalized communities, filling gaps in the transportation network and addressing latent travel demand [9–12], but no work has shown how this plays out in more naturalistic experiments.

To examine these trends, in 2022, researchers at the University of San Francisco (USF) launched a pilot "Research Rider" program with Cruise LLC ("Cruise") to better understand how an AV service can fill transportation needs for USF students. The goal of the program was to establish a partnership between industry and academia, to address lingering transportation policy challenges in the city, including transit deserts, congestion, and increased sustainable modes of transport. Cruise was permitted to service areas in close proximity to the USF campus during off-peak/late night hours, when transit served students least, initially offering free rides between the hours of 11:00pm and 5:00am in the Operational Design Domain (ODD) approved by the California Department of Motor Vehicles (DMV) and California Public Utilities Commission (CPUC).

This paper explores how individuals use autonomous vehicles, and how their travel behaviors change when provided access to new autonomous services during their involvement in the early access rider program. This research is particularly important since Cruise recently received full approval from the California Department of Motor Vehicles and the California Public Utilities Commission to operate a paid driverless ridesharing service in a limited ODD within San Francisco, and other vendors such as Waymo may undertake similar driverless operations in the future. Given these developments, this paper seeks to: 1) explore travel choices and transport policy challenges related to AVs; and to 2) evaluate the customer experience and business model for operational competitiveness.

First, a background and literature review are provided, offering a glimpse at how many transportation researchers have attempted to model or simulate travel behavior in the absence of a fully functional driverless framework for passenger testing. We then lay out a naturalistic experiment where roughly 250 participants were offered early access to free rides in autonomous vehicles. Following presentation of the results, a discussion on the implications of this work ensues, along with implications for future research.

2 Background and Literature Review

Since the early 2010s, planners and engineers have speculated about how the future of fleet-based mobility, rideshare technology, and automation will impact urban transportation. Much of this early research dealt with casual carpool programs and mobility-as-a-service platforms, that led to the emergence of app-based rideshare services such as Uber and Lyft [13–15]. But more recently, efforts have been made to map the evolution of rideshare/transportation network companies to predict the future inclusion of autonomous vehicles [2, 3, 16–18].

While the accuracy of this predicted evolution remains to be seen, the trend toward increasingly sophisticated vehicles that can drive without significant passenger supervision, pose many interesting questions. Researchers have predicted both increases in vehicle miles travelled (VMT) due to greater mobility access for underserved populations, and decreases in greenhouse gas emissions [1, 19, 20]. Others have anticipated increased roadway capacity and less need for parking or private vehicles in a pivot to a fleet-owned model [5, 21–23].

AVs have the potential to change the way we view transportation. There are opportunities to increase space for cyclists, pedestrians, and other vulnerable road users as well as to better meet the needs of those who require more accessible rides - including for non-ambulatory, blind/low vision, and deaf/hard of hearing individuals. [6, 24, 25]. Regulators have called for increased advanced planning to prevent negative externalities and to tailor deployments to ensure more equitable and sustainable access across urban, suburban, and rural environments. [26–30].

Still, questions remain, particularly around how planners can think of autonomous platforms as complementary to existing transit services. An ample body of work deals with travel mode choices and programs to promote walking, cycling and transit use [31–34], but little is known about the impacts of automation on travel behavior - particularly in how it can complement transit during off-peaks service hours. Limited work explores synergies with transit in ridesharing environments [9, 10, 13, 35] and found that in most cases rideshare both complements (reducing network gaps/last mile issues) as well as supplements (serving when service is limited or unreliable) transit. No work thus far has looked at actual rider behavior in an autonomous platform, how it relates to transit, and what business models best serve AV fleet operators as they align service design with societal values to provide sustainable and equitable transportation systems. This research fills that gap in knowledge.

3 Methods

To conduct this work, USF researchers, collaborated with the Cruise Policy Research and Ridehail Business Unit teams to design a project in which USF students would be given early access to Cruise's all-electric self-driving fleet to take rides to and from select geographies within the allowed hours (originally 11:00 PM and 5:00 AM and thereafter extended to the hours between 10:30 PM and 5:00 AM). The joint team felt that the project presented a unique opportunity to shape the future of autonomous urban mobility, at its outset, by engaging an initial rider base that could see the greatest benefit from this service due to user travel behavior and the relative lack of reliable and available transportation options within the approved ODD. An initial "screener" survey was launched by the USF team in March 2022 to determine project eligibility.

Roughly 250 students responded with an interest in participating in the Research Rider program. Baseline information was collected including student travel patterns, behavioral preferences, residential location, and basic demographics. Critical demographic attributes included: 39% white; 61% nonwhite; 57% of white riders' primary method of evening transportation was rideshare; 63% of nonwhite riders' primary method of evening transportation was rideshare; 14% of white riders' primary method of evening transportation was public transit; 18% of nonwhite riders' primary method of evening transportation was public transit and 38% live on campus in San Francisco.

As shown in Fig. 1 most travel during the day was for school, however 73% of trips after hours were for social or recreational purposes. Students indicated that the bulk of

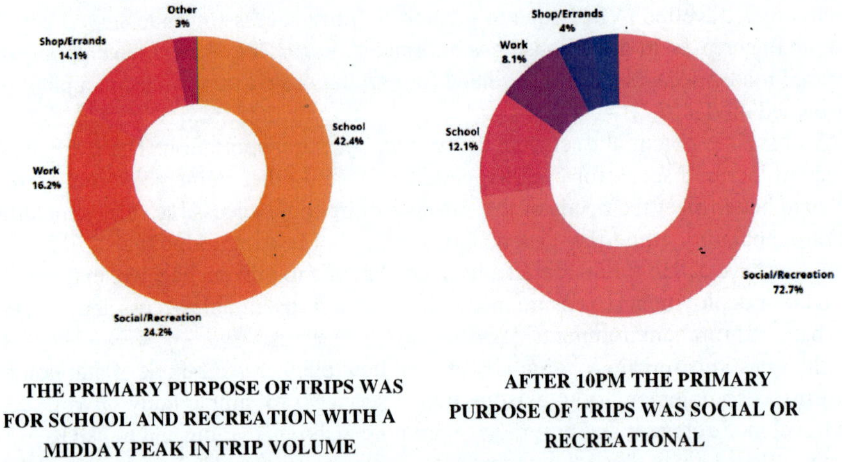

Fig. 1. Variation in trip purpose between the day and evening hours.

this travel (42%) was done via transit during the day, but after-hours transit was far less used with students instead relying on rideshare and driving (See Fig. 2). Most students felt that transit connectivity, trip distribution and service access were important in what they would expect from an AV ride.

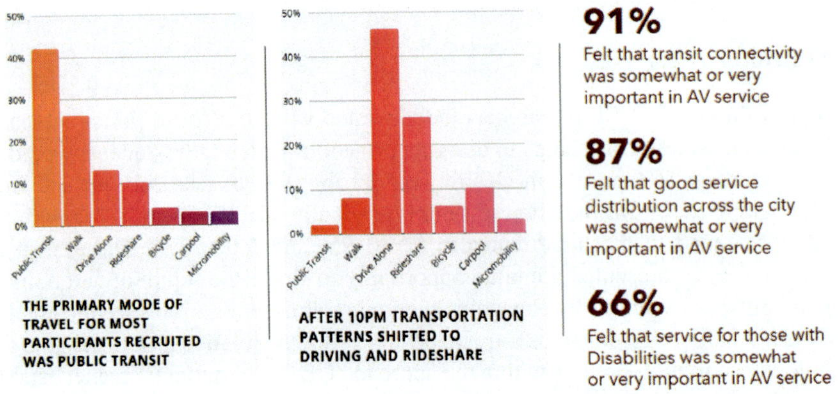

Fig. 2. Variance in trip mode between the day and evening hours, and basic consumer preferences.

These baseline indicators were used to create a statistically significant cross-section of approximately 160 students (representing diversity with regards to the demographics, location, availability, and travel behavior) who were invited to join the Research Rider pilot program. These students were encouraged to download the Cruise app and asked to take a survey each week documenting their trips as a part of the program. Trips were made in the ODD located in the western part of San Francisco, as shown in Fig. 3. Which was approved for fully driverless rideshare services in early 2022.

Fig. 3. Cruise Initial ODD Approved by CPUC for fared and driverless service. Source: Cruise

47 of the initial 160 students took rides. Each week they were asked to report their rides in a post-ride survey delivered to their cell phone or email address. The survey documented pick and drop-off location, trip purpose, time of day, and how/if the autonomous trip would have otherwise been made.

It is worth noting that these results do reflect a strong degree of self-selection in that students not only elected to participate in the survey but to take and then report AV rides. While this represents a limitation of the work, in that student riders deliberately chose AVs as their preferred alternative means of travel, it could also be cast as a strength in terms of understanding user profiles, preferences and trends. Furthermore, this does not wholly invalidate results in that the attributes and use patterns of those who self-select AVs for use were precisely the point of the research. In addition, our student population represents a demographic that is often underserved in terms of availability of transit options, especially during off-peak hours, which could give a valid representation of the transit challenges faced by many other younger or older populations within "transit deserts". Additionally, while it could be argued that our work is limited in that rides were offered for free (as Cruise was not approved for fared rides at the time of our work), we believe that by removing cost as a factor it allows us to isolate behavioural and service efficiencies independently of pricing economics.

Results were tabulated and summarized and select riders and non-riders were invited to focus groups with the research team to discuss their experiences.

4 Results

Of the 47 riders, 50 rides were reported. When looking at how they expected to travel vs. how they actually traveled some interesting themes emerge. As shown in Fig. 4, the largest use case for riders in the pilot program were social/recreational trips. In the initial screener survey, the cohort that ended up being accepted as riders put their expected usage of AVs for social/recreation at 70.2% of trips. The actual results were very close to this at 71.4% of trips being for social/reaction purposes. This result was expected for many reasons, most notably given the nature of the pilot program operating hours (11 PM–5 AM).

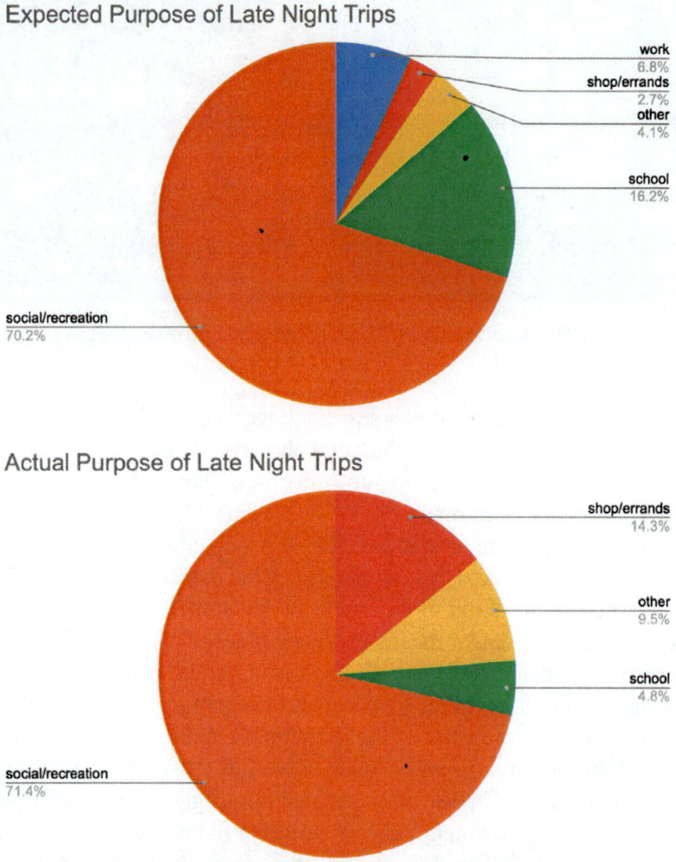

Fig. 4. Expected vs. Actual AV Ride Purpose.

Somewhat unexpectedly however, were the larger number of utilitarian trips for the purpose of shopping/errands. Initially, these trips were expected to account for less than 3% of total trips. Instead, they ended up accounting for 14.3% of trips—a substantially larger amount than anticipated. In one of the focus groups, a participant described how the Cruise pilot program made it possible for her to "go do her laundry at her work" (less expensive than finding another option). Another described an experience of how the AVs were vital in assisting them to go to get late night food after hours because "(their) neighborhood shuts down" - both these examples speak to the utility and cost savings that the availability of these vehicles can provide to communities, even recognized via such a targeted initial launch.

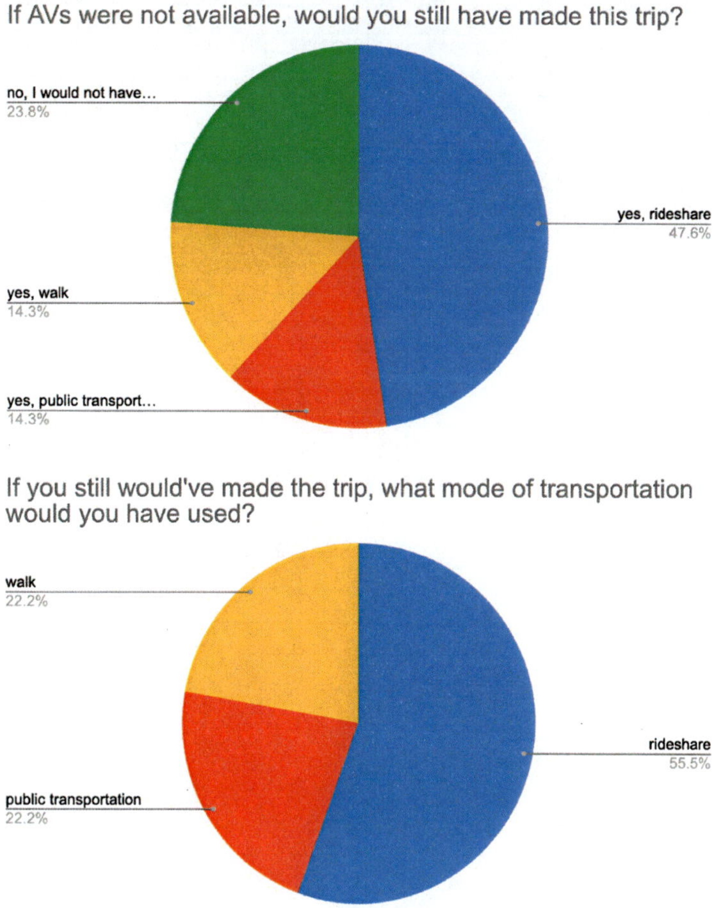

Fig. 5. Expected vs. Actual AV Ride Purpose.

Results also indicated that the largest portion of trips being taken through this pilot program would have been made irrespective of whether the Cruise vehicles had been available or not (see Fig. 5). More than three quarters of trips (76.2%) would have occurred even if AVs were not available representing the existing travel demand where the Cruise AV was an acting substitute. Only 23.8% of trips were that would not have been otherwise, which represents either latent demand (existing desire for travel that is underserved by the current transport network) or induced demand (travel generated as a result increased availability of transportation resources).

In sum whether or not the trips represent latent or induced demand they do show a general increase in total travel on the road but an increase that is more modest than many academics have speculated, and may serve critical needs not served by public transport

Table 1. Rider/Non-Rider Mode of Transport for all Trips (Note: not representative of total reported trips)

Rider/Non Rider v. Late Night Mode of Transport	Bicycle	%	Drive Alone	%	Carpool	%	Micro-mobility	%	Public Transit	%	Ride-share	%	Walk	%	Other	%	Grand Total
rider	**0**	**0%**	**0**	**0%**	**2**	**11%**	**1**	**6%**	**3**	**17%**	**10**	**56%**	**1**	**6%**	**1**	**6%**	**18**
% in category	0%		0%		10%		17%		6%		9%		10%		4%		7%
nonrider	**5**	**2%**	**17**	**8%**	**18**	**8%**	**5**	**2%**	**48**	**21%**	**98**	**44%**	**9**	**4%**	**25**	**11%**	**225**
% in category	100%		100%		90%		83%		94%		91%		90%		96%		92%
Grand Total	**5**	**2%**	**17**	**7%**	**20**	**8%**	**6**	**2%**	**51**	**21%**	**108**	**44%**	**10**	**4%**	**25**	**10%**	**245**

systems—for example for school, work or errands. Of the trips that would have been made regardless (76.2%), more than half or 55.5% would have been made through ridesharing services like Uber and Lyft. However, somewhat interestingly, amongst the trips that would not have been made otherwise (induced demand) 75% were for shopping/errand related trips.

In comparison to individuals that filled out the screener survey but ultimately did not participate in the pilot program (non-riders), riders (those who filled out the screener and then ultimately did participate) differed in two interesting ways. First, riders were slightly more likely to have their primary nighttime mode of transportation be rideshare than non-riders. As shown in Table 1, 56% (55.5%) of riders' primary nighttime transportation was rideshare, while services like Uber and Lyft only made up 44% of non-riders. Second, riders were slightly less likely to have primary nighttime transport be public transit compared to non-riders. 17% of riders' primary nighttime mode of transport was public transportation verses 21% of non-riders who were slightly more likely to take transit.

Amongst riders, a few factors were highlighted as especially important to encourage future AV usage. As illustrated in Fig. 6, the most important factor related to AV use was cost and the potential opportunity for AVs to be less expensive than alternatives (which for most riders is ridesharing). However, in the survey responses and focus group dialogues, themes of reliability and convenience also emerged and were prevalent. One participant mentioned pick up and drop off as particularly important. He described a scenario where he had an issue calling a vehicle to a particular spot and had to walk several blocks uphill to catch his ride.

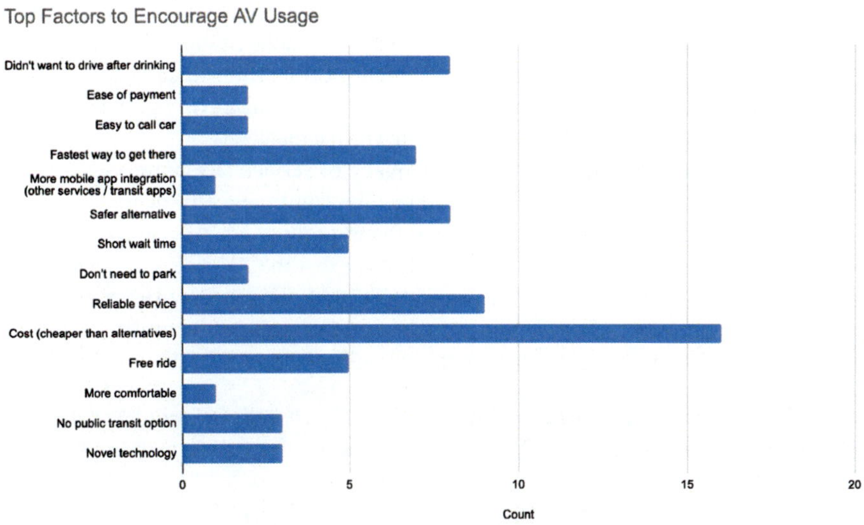

Fig. 6. Factors Influencing a Ride in an AV.

5 Discussion

Although the expansiveness of our rider pilot data is limited, our results suggest that afterhours AV service can serve as both a complement and supplement to rideshare and transit, and fill needs for latent travel demands—particularly in service locations or during times where transit is not as frequent. This is important because of the vast service equity implications of having AVs in these locations and at these times, particularly countercyclically to existing transit availability. A vast body of literature has shown that many transit networks have gaps that have historically marginalized the most vulnerable communities. Our work illustrates that AVs can be part of the solution to beginning to fill these service gaps.

Of equal importance is our finding that many of the trips that were taken by riders would have occurred anyway (76.2%). This is an important finding as it provides a counter-narrative to frequent speculation that AVs will result in a significant amount of induced demand. In our Pilot Project, the results show the opposite. The bulk of AV trips (77.7%) were substitutions for rideshare and then transit—trips or modes that likely may have been inefficient, underserved or less adequate. For the 23.8% of trips that respondents said they "would not have made otherwise" we would argue against the narrative that this represents induced demand, first because some of these trips were simply related to the novelty of riding in a new, high-tech self-driving car, and second because of the increase in spatial-temporal accessibility these vehicles provided. It is very clear that many of the young adults in our study (many of whom did not own cars) may have been under-resourced in terms of their available modes for travel (particularly during the service hours) and that their travel needs had been constrained by this limitation. Many had the desire to travel but were under-resourced in their ability to do so. Although there may be some inducement of travel, this clearly represents a large share of latent demand, and in terms of increase in total travel represents a total increase in travel below what some academics have speculated.

Moreover, it is worth noting that the inadequacy of service (for both rideshare and transit) was linked to two factors: safety and service convenience/frequency. We documented many situations, particularly in our focus groups, where our female riders expressed safety concerns. Many stated that they felt that the AV ride was safer—even in a shared environment. Riders were comfortable with the current AV safety protocols and features in place. Focus group participants would also complain about the convenience of bus routes and then praise the reliability of the AV, saying things like:

> *"Sometimes getting out to where I live there's not a whole lot of real good direct bus service.... And then certainly, the fact that the Cruise vehicles are available is pretty darn awesome..."*

With regard to safety many women who were interviewed expressed that the capability of riding in an autonomous rideshare vehicle was a vast improvement over traditional rideshare (e.g. Uber/Lyft). Many of these riders said they preferred riding without a driver during late-night hours since they had experienced or heard of issues of driver aggression or assaults. For example, in one focus group a young woman described a situation where a driver had waited outside of her home after dropping her off which made her feel very uncomfortable. She felt that she did not have to worry about this with an AV.

Given these performance factors, it is clear that some of the mode substitution behavior might be a result of better and more reliable service, but individuals also identified the importance of additional service factors, such as the environmental impact/sustainability of the ride as well as comfort. This also matched up well with consumer preference data from the original screener surveys. In those snapshots, as shown in Fig. 7, 87% of potential riders thought good distribution of service was very important and 91% saw transit integration/connectivity as either very or somewhat important. Sixty-six percent (66%) felt the same way about disabled access and better service for those with disabilities—including those with reduced mobility or low vision/hearing impairments.

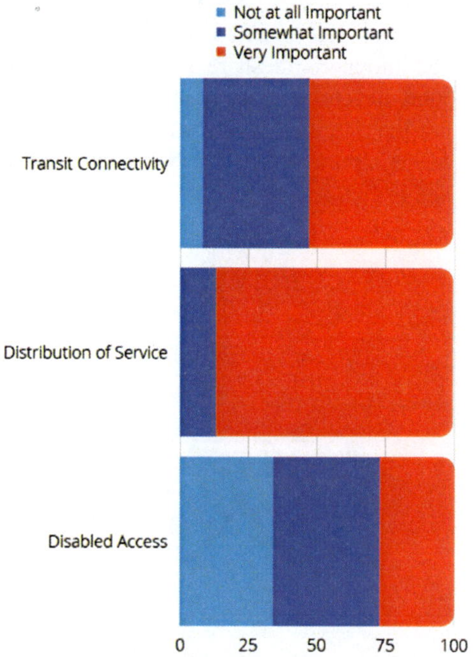

Fig. 7. Rider Preferences for AV Rides.

Many emphasized the positive elements of the vehicles and expressed a desire for more hours of service; beyond the late-night hours, and an expansion beyond the fairly restricted locations/ODD available in the rider pilot. These time and geographic restrictions are important to emphasize because our survey and focus group participants did not realize the large role regulators play in determining where/when vehicles can operate. Clearly many were frustrated with the current transit service quality and our research suggests that there is demand for access to a larger ODD in San Francisco (and beyond) and extended hours that could allow AVs to further supplement existing transit. These emergent platforms, that are purpose built for sharing and meeting the needs of diverse communities and those with disabilities, could provide complementary services that enhance existing high-capacity transit.

Likewise, there was equal frustration in focus groups about drop-off situations where vacant driveways appeared to be available for curbside passenger disembarkation, but the AV continued in search of vacant curb space further from the desired destination. This is a policy issue that needs focus in future discussion since the California vehicle code (and potentially other locations) does not permit drop off at vacant driveways but allows commercial vehicles to double park to permit the safe exit of passengers. This is often confused by the media; however, it is an opportunity for regulatory dialogue in the future - particularly in light of recent efforts advanced by jurisdictions like the City of San Francisco's Curb Management Strategy.

Finally, while our work is limited in that it did not engage with the cost of an autonomous ride as compared to another type of ride, since our preliminary pilot predated the introduction of priced rides, our focus was more exploratory in nature to investigate how cost played into these rideshare decisions as well as other factors. As is illustrated in Fig. 8 we found that cost was one of many "pull" factors that attracted pilot riders to the program—not the only factor. In many cases factors related to reliability, convenience and safety. More importantly we did investigate willingness to pay as a part of our work as is illustrated in Fig. 8.

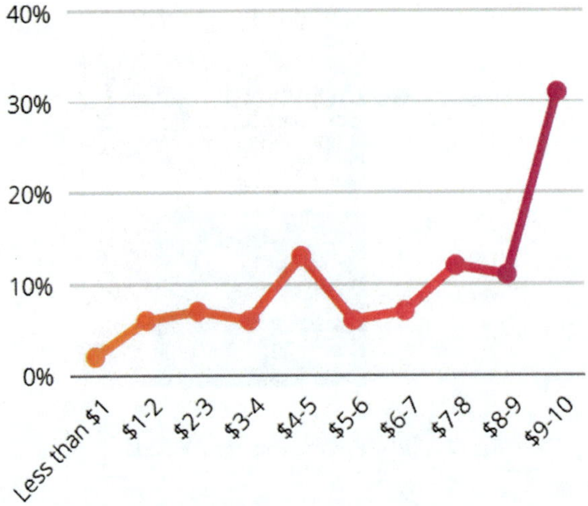

Fig. 8. Willingness to Pay for Autonomous Ride

On average participants were likely to pay approximately $10 for a short ride, and based on preliminary introduction of pricing (priced rides that were introduced by Cruise after this pilot) rates were lower than both traditional rideshare or transit. More than 30% were willing to pay between $9 and $10. This factor warrants future investigation since it may offer potential cost/pricing efficiencies, new ways of thinking about service delivery during off-peak hours and new methods of providing feeder lines to trunk transit service.

6 Conclusions

This paper evaluates a first-of-its-kind program, offering passengers autonomous rides in Cruise vehicles between the hours of 11:00pm–5:00am when traditional transit services are less prevalent. The aim of this project was to better understand how an AV service can fill transportation needs for USF students acting as a representative population of potential initial riders – particularly as those who may not have as robust access to existing transportation solutions. Results indicate that more than 76% of reported travel by AV riders was mode substitution, largely from rideshare (at 55.5%) and transit (at 22.2%).

Of the 55.5% of trips that replaced rideshare travel—most made were for recreational and commercial purposes such as errands and shopping. These results suggest that most AV trips offer an opportunity to address network gaps and last mile connectivity. Most trips replaced existing travel already on roadway network, and arguably made it more efficient, and the trips that were new (23.8%) appeared to meet existing needs underserved by current services. Individuals largely enjoyed their experience in AVs despite constraints on how and when the vehicles can operate, and many riders noted a desire for more areas of service. These findings are significant as new shared vehicle solutions emerge in the marketplace, and our research suggests that there is demand for access to a larger ODD in San Francisco (and beyond) and extended hours that could compliment existing high-capabity transit routes. In the future, when less restrictions exist on AV operations, more studies should be conducted to see the impact of AV introduction on existing transit systems as well as the efficiency of the street grid network itself [36].

While the pilot project involved a relatively small sample, had concentrated focus on students and predated the ability to introduce priced/paid rides, it offers a window into travel behavior and mode substitution for actual AV rides. This is import because despite these limitations, the study provides information on how trips might change, and how users might benefit from the availability of AVs during hours where traditional transit is limited. With this goal in mind the project provides a potential look at how AVs could potentially provide better service and increased mobility to traditionally underserved passengers and communities.

The paper and project also opens up numerous pathways for further research. Understanding the ways in which AVs will be used in cities is essential as their adoption is beginning to move from research to commercial offerings. Cruise has recently received full approval from the California Department of Motor Vehicles and the California Public Utilities Commission to operate a paid driverless ridesharing service in a limited ODD in San Francisco. Other vendors such as Waymo are offering drivered ridesharing service in San Francisco and may provide autonomous rideshare services in the future. These developments in the marketplace lead us to believe that these types of transport services hold promise for the future of how we think about transportation systems and eliminating service deserts.

More importantly this work opens the aperture as to how public and private vendors might work in concert to improve user experiences and public transport service. There is room in the future to investigate the possible opportunities for integration of AVs to contribute to a shared public and private transit service. In this light, it is essential that cross-sector collaboration dialogues continue as this technology continues to advance.

This will ensure that future AV services most aptly compliment and supplement existing modes of transportation as they are deployed.

References

1. Fagnant, D.J., Kockelman, K.M.: The travel and environmental implications of shared autonomous vehicles, using agent-based model scenarios. Transport. Res. Part C Emerg. Technol. **40**, 1–13 (2014)
2. Litman, T.: Autonomous Vehicle Implementation Predictions. Victoria Transport Policy Institute, vol. 28 (2014)
3. Harb, M., Xiao, Y., Circella, G., Mokhtarian, P.L., Walker, J.L.: Projecting travelers into a world of self-driving vehicles: estimating travel behavior implications via a naturalistic experiment. Transportation **45**(6), 1671–1685 (2018)
4. Batur, I., et al.: Understanding interest in personal ownership and use of autonomous vehicles for running errands: an exploration using a joint model incorporating attitudinal constructs. Transport. Res. Rec. 03611981221107643 (2022). https://doi.org/10.1177/03611981221107643
5. Riggs, W., Beiker, S.A.: Business models for shared and autonomous mobility. In: Meyer, G., Beiker, S. (eds.) AVS 2019. LNM, pp. 33–48. Springer, Cham (2020). https://doi.org/10.1007/978-3-030-52840-9_4
6. Riggs, W., Pande, A.: Gaps and opportunities in accessibility policy for autonomous vehicles. Publication 2106. Mineta Transportation Institute (2021)
7. Westervelt, M., et al.: UpRouted: Exploring Microtransit in the United States (2018)
8. Lazo. Uber, Lyft Partner with Transportation Authority to Offer Paratransit Customers Service in Boston. Washington Post (2016)
9. Gehrke, S.R., Felix, A., Reardon, T.G.: Substitution of ride-hailing services for more sustainable travel options in the greater Boston region. Transport. Res. Rec. 0361198118821903 (2019). https://doi.org/10.1177/0361198118821903
10. Clewlow, R.R., Mishra, G.S.: Disruptive transportation: the adoption, utilization, and impacts of ride-hailing in the United States. Publication UCD-ITS-RR-17-07. University of California, Davis, Institute of Transportation Studies, Davis, CA (2017)
11. Brown, A.E.: Ridehail Revolution: Ridehail Travel and Equity in Los Angeles. UCLA (2018)
12. Leistner, D.L., Steiner, R.L.: Uber for seniors?: exploring transportation options for the future. Transp. Res. Rec. **2660**(1), 22–29 (2017). https://doi.org/10.3141/2660-04
13. Rayle, L., Dai, D., Chan, N., Cervero, R., Shaheen, S.: Just a better taxi? A survey-based comparison of taxis, transit, and ridesourcing services in San Francisco. Transp. Policy **45**, 168–178 (2016). https://doi.org/10.1016/j.tranpol.2015.10.004
14. Riggs, W.: Disruptive Transport: Driverless Cars, Transport Innovation and the Sustainable City of Tomorrow. Routledge, London (2019)
15. Shaheen, S.A., Cohen, A.P.: Carsharing and personal vehicle services: worldwide market developments and emerging trends. Int. J. Sustain. Transp. **7**(1), 5–34 (2013)
16. Alemi, F., Circella, G., Handy, S., Mokhtarian, P.: What influences travelers to use uber? Exploring the factors affecting the adoption of on-demand ride services in California. Travel Behav. Soc. **13**, 88–104 (2018). https://doi.org/10.1016/j.tbs.2018.06.002
17. Greenblatt, J.B., Shaheen, S.: Automated vehicles, on-demand mobility, and environmental impacts. Curr. Sustain./Renew. Energy Rep. **2**(3), 74–81 (2015)
18. Grush, B., Niles, J.: The End of Driving: Transportation Systems and Public Policy Planning for Autonomous Vehicles. Elsevier, Amsterdam (2018)

19. Duell, M., Levin, M.W., Boyles, S.D., Waller, S.T.: Impact of autonomous vehicles on traffic management: case of dynamic lane reversal. Transp. Res. Rec. **2567**(1), 87–94 (2016). https://doi.org/10.3141/2567-10
20. Greenblatt, J.B., Saxena, S.: Autonomous taxis could greatly reduce greenhouse-gas emissions of US light-duty vehicles. Nat. Clim. Chang. **5**(9), 860–863 (2015). https://doi.org/10.1038/nclimate2685
21. Appleyard, B., Riggs, W.: Doing the Right Things" Before "Doing Things Right": A Conceptual Transportation/Land Use Framework for Livability, Sustainability, and Equity in the Era of Autonomous Vehicles. Washington, D.C. (2018)
22. Zhang, W., Guhathakurta, S., Fang, J., Zhang, G.: Exploring the impact of shared autonomous vehicles on urban parking demand: an agent-based simulation approach. Sustain. Cities Soc. **19**, 34–45 (2015). https://doi.org/10.1016/j.scs.2015.07.006
23. Zhang, W., Guhathakurta, S.: Parking spaces in the age of shared autonomous vehicles: how much parking will we need and where? Transport. Res. Rec. J. Transport. Res. Board **2651**, 80–91 (2017)
24. Riggs, W.: End of the Road: Reimagining the Street As the Heart of the City. Bristol University Press, Bristol (2022)
25. Riggs, W., Appleyard, B., Johnson, M.: A design framework for livable streets in the era of autonomous vehicles. Urban Plann. Transp. Res. (2020)
26. Bansal, P., Kockelman, K.M., Singh, A.: Assessing public opinions of and interest in new vehicle technologies: an Austin perspective. Transport. Res. Part C Emerg. Technol. **67**, 1–14 (2016). https://doi.org/10.1016/j.trc.2016.01.019
27. Crute, J., Riggs, W., Chapin, T., Stevens, L.: Planning for Autonomous Mobility. Publication PAS 592. American Planning Association, Washington D.C. (2018)
28. Guerra, E.: Planning for cars that drive themselves metropolitan planning organizations, regional transportation plans, and autonomous vehicles. J. Plann. Educ. Res. 0739456X15613591 (2015). https://doi.org/10.1177/0739456X15613591
29. Riggs, W., Larco, N., Tierney, G., Ruhl, M., Karlin-Resnick, J., Rodier, C.: Autonomous vehicles and the built environment: exploring the impacts on different urban contexts. In: Meyer, G., Beiker, S. (eds.) AUVSI 2018. LNM, pp. 221–232. Springer, Cham (2019). https://doi.org/10.1007/978-3-319-94896-6_19
30. Wellik, T., Kockelman, K.: Anticipating land-use impacts of self-driving vehicles in the Austin, Texas, region. J. Transp. Land Use **13**(1), 185–205 (2020). https://doi.org/10.5198/jtlu.2020.1717
31. Pinjari, A.R., Pendyala, R.M., Bhat, C.R., Waddell, P.: Modeling the Choice Continuum: Integrated Model of Residential Location, Automobile Ownership, Bicycle Ownership, and Commute Tour Mode Choice Decisions (2008)
32. Buehler, R., Pucher, J., Bauman, A.: Physical activity from walking and cycling for daily travel in the United States, 2001–2017: demographic, socioeconomic, and geographic variation. J. Transp. Health **16**, 100811 (2020). https://doi.org/10.1016/j.jth.2019.100811
33. Pucher, J., Dill, J., Handy, S.: Infrastructure, programs, and policies to increase bicycling: an international review. Prev. Med. **50**(Supplement), S106–S125 (2010). https://doi.org/10.1016/j.ypmed.2009.07.028
34. Riggs, W.: Testing personalized outreach as an effective TDM measure. Transport. Res. Part A Policy Pract. **78**, 178–186 (2015). https://doi.org/10.1016/j.tra.2015.05.012
35. Hall, J.D., Palsson, C., Price, J.: Is uber a substitute or complement for public transit? J. Urban Econ. **108**, 36–50 (2018). https://doi.org/10.1016/j.jue.2018.09.003
36. Boeing, G., (Billy) Riggs, W.: Converting one-way streets to two-way streets to improve transportation network efficiency and reduce vehicle distance travelled. J. Plann. Educ. Res. (2022). https://doi.org/10.1177/0739456X221106334

Part III: Vehicle Technology Development and Testing

Surrogate Measures of Automated Vehicle Safety

Erik K. Antonsson(✉)

Streetscope, Inc., 130 W. Union St., Pasadena, CA 91103, USA
erik@streetscope.com

Abstract. Surrogate measures of traffic safety replace collision statistics as a means of assessing the safety of a roadway, intersection, vehicle, or mobility system. Effective and consistent surrogate measures of traffic risk and safety that will be useful to ADS stakeholders — including AV developers, traffic infrastructure developers and managers, regulators, legislators, and the public — will have a number of essential characteristics, including *monotonicity* and *scalability*.

Trailing indicators, such as collision statistics, are a poor methodology for improving safety, In addition, the use of trailing indicators incurs pain and loss on society, and is not an ethically acceptable approach. *Leading* indicators, based on non-collision interactions, include: Traffic Conflicts, Time-to-Collision (TTC), Post-Encroachment Time (PET), Instantaneous Safety Metric (ISM), harsh accelerations and turns (generally measured by an inertial measurement unit (IMU)), AV Control System Disengagements, and near-misses or near-crash events.

Surrogate measures, reviewed here, gather, process, and in some cases predict traffic movement, or control system behavior, and produce a (sometimes quantitative) score reflecting the riskiness or safeness of the behavior of vehicles in traffic.

Nomenclature

ADS	Automated Driving System [1]
AV	Automated Vehicle; Autonomous Vehicle
IMU	Inertial Measurement Unit
ISM	Instantaneous Safety Metric [2]
MPrISM	Model Predictive Instantaneous Safety Metric [3]
PET	Post-Encroachment Time [4]
RSS	Responsibility-Sensitive Safety [5]
TCT	Traffic Conflicts Technique [6]
TTC	Time-to-Collision [7]
NHTSA	National Highway Traffic Safety Administration, U.S. Department of Transportation

G. Meyer and S. Beiker (Eds.): ARTSymposium 2022, LNMOB, pp. 111–132, 2023.
https://doi.org/10.1007/978-3-031-34757-3_10

1 Introduction

1.1 Background

Evaluation of the risk and safety of the operation of vehicles in traffic is critical to the deployment and public acceptance of automated mobility systems.

Historical collision statistics are important benchmarks; however, they are problematical for assessing the risk and safety of new systems, either roadways, traffic controls, vehicles, or mobility systems. This is precisely because such trailing or lagging indicators of traffic safety require the accumulation of collision statistics and necessarily therefore require collisions to occur [6,8,9]. Traffic collisions incur property damage, injuries, and death, and the ethical problems of inflicting pain and loss on society (in the process of determining the safety of an automated mobility vehicle or system) are unacceptable.

Because of the problems with lagging indicators, *leading* indicators or measures of the risk and safety of the operation of vehicles in traffic are required, and are commonly referred to as "surrogate" measures of traffic safety.

Surrogate measures of traffic safety have been evaluated and compared previously, including: by the U.S. Federal Highway Administration [10,11], Transportation Research Board [12], U.S. National Highway Traffic Safety Administration [13], University of Toronto [14], Rutgers University [15], Canadian Journal of Civil Engineering [16], Riga Technical University [17], Transport Reviews [18], Center for Road Safety at Purdue University [19], Lund University [20], International Encyclopedia of Transportation [21], Czech Republic Transport Research Centre [22], University of California, Berkeley [23], and *Accident Analysis & Prevention* [24].

1.2 Safety Assessment

Safety measurements, whether they are lagging indicators based on collision statistics or leading surrogate measures, work within a broader process of safety assessment. As a general matter, the process of assessing performance (including safety) requires three distinct elements:

Testing: One or more tests that challenge the performance which is to be assessed.

Evaluation: One or more ways that the results of the test are measured.

Reference Standard: One or more standards that are used as a reference against which to compare the measured performance.

For example, in assessing a student's understanding:

Testing: A test would be devised consisting of a series of questions for the student to answer.

Evaluation: The student's responses would be evaluated to determine whether they are correct. The evaluation of the student's responses would be aggregated into an overall measure or score, *e.g.*, 92 out of 100.

Reference Standard: The student's evaluated score would be compared to other students' scores to determine an overall result to be reported on the student's academic record.

All three elements must be used to assess the performance of an individual or system under test. Testing alone is not sufficient. Evaluation of the test responses, and comparison of the evaluation of the responses with a reference standard, are both required to assess the performance.

While there is considerable active discussion relating to testing of ADS (whether in simulation by means of scenarios, or on-track testing, or on-road testing with traffic), evaluation of the results of such tests are frequently given considerably less attention or omitted altogether. The absence of one or more consistent evaluations has produced the present environment, where reference standards have not yet been developed or promulgated to actually gauge whether the control and behavior of automated vehicles is sufficiently safe to operate on public streets and roads.

1.3 Measurements and Assessments

A number of (non-collision) surrogate measures exist, starting with the Traffic Conflicts Technique (TCT), introduced by Perkins and Harris at General Motors Research in 1968 [6,25] and followed shortly thereafter by Time-to-Collision (TTC) introduced by Hayward in 1972 [7]. Since that time, a variety of additional traffic safety measures have been proposed and used, including both leading and trailing indicators [26]. This chapter reviews the existing methods, and evaluates each against a set of characteristics needed for a measure to be consistent and effective. This chapter also briefly reviews a novel leading measure of collision hazards in traffic that overcomes difficulties with prior measures, and is quantitative, objective, continuous, and general.

For the introduction of new mobility systems, particularly automated vehicles, *leading* indicators of traffic risk and safety must be used. However, existing leading road safety indicators have limitations and/or shortcomings, and a new general measure is needed. A summary of the most critical shortcomings of the most commonly used surrogate measures is presented below.

Importantly, a general quantitative measure of traffic safety and risk must treat each traffic object (vehicle, pedestrian, bicyclist, *etc.*) as a "black box", meaning that only the external behavior of the object can be utilized in computing the measure. This is identical to the conditions of an on-road test of a human driver, where the evaluator simply observes the actions (behaviors) of the vehicle as it is being driven, and is not engaged in a dialog with the driver about what he/she sees nor what considerations she/he is making regarding control actions.

For the assessment of a human driver, the same three elements are required:

Testing: A test which challenges the driver's skills, including, *e.g.*, starting, turning, stopping, lane changing, merging, parking.

Evaluation: The student's actions and responses are evaluated by the evaluator to determine whether they are in compliance with local traffic laws and regulations and are safe, *e.g.*, speeds, spacing, maneuvers.

Reference Standard: The evaluator uses his training and experience to judge whether the driver being tested has demonstrated sufficient mastery and safe driving behavior to be granted a license to drive in traffic.

In the case of the assessment of human drivers, the consideration and evaluation of how safely the driver operated his/her vehicle, and how safe the driver will be in traffic once licensed, are entirely qualitative (except in the unusual case where the driver is involved in a collision during the test).

In the case of ADS, qualitative judgments of observers will not suffice to allow automated vehicles to operate in traffic. The public, regulators, legislators, and insurers, have made clear that automated machines operating in life-safety critical applications, such as vehicles in traffic, will require consistent and effective quantitative assessment of safety, and must demonstrate safe behavior at least at the level of existing human drivers (and preferably a greater level of safety).

2 Near-Misses; Near-Crash Events

Hayward suggested that near-misses could be an effective leading indicator of traffic risk and safety:

> NEAR-MISS traffic events have been considered for use as predictors of accident rate characteristics at roadway locations. The near miss, loosely defined, is a traffic event that produces more than an ordinary amount of danger to the drivers and passengers involved. Near misses would appear to be closely related to the accident pattern witnessed at a location and, therefore, could become an attractive alternative measure to accident-based safety determination.
> [7, page 24]

Other fields have used near-misses as leading indicators of risk for some time [27–29], by means of self-reporting that has been shown to reduce the instances and severity of incidents and losses. For example, near-misses have been collected and analyzed in civil aviation since at least 1958:

> The Aviation Safety Reporting System, or ASRS, is the US Federal Aviation Administration's (FAA) voluntary confidential reporting system that allows pilots and other aviation professionals to confidentially report near misses or close call events in the interest of improving aviation safety. [30]

The chemical processing industry has also implemented near-miss management systems.

> In review of adverse incidents in the [chemical] process industries, it is observed, and has become accepted, that for every serious accident, a larger number of incidents result in limited impact and an even larger number of incidents result in no loss or damage.
> ...
> Despite their limited impact, near misses provide insight into accidents that could happen.
> [31, page 445]

"Near-crash" events were explored in the 2006 DOT/NHTSA SHRP2 100-Car Naturalistic Driving Study:

> • Near-Crash: Any circumstance that requires a rapid, evasive maneuver by the subject vehicle, or by any other vehicle, pedestrian, cyclist, or animal, to avoid a crash. A rapid, evasive maneuver is defined as steering, braking, accelerating, or any combination of control inputs that approaches the limits of the vehicle capabilities. As a guide, a subject vehicle braking greater than 0.5 g or steering input that results in a lateral acceleration greater than 0.4 g to avoid a crash, constitutes a rapid maneuver.
>
> As shown, while these criteria were based somewhat upon quantitative kinematic criteria, they were subjective in nature. While such definitions were useful for purposes such as classifying video data, they were not useful for precisely defining events or as criteria for other purposes, such as warning algorithms.
> [32, Page 139]

> [N]ear-crashes, since they (by definition) have many of the same elements as a crash, may provide useful insight into the risk associated with driver behavior and environmental factors in combination with crashes. This ... benefit, if it can be validated, can provide a powerful tool for analyzing naturalistic driving data since near-crashes occur at a rate of roughly 10 to 15 times more frequently than crashes. Thus, there is a need to better understand the relationship between crashes and near-crashes as well as the impact of using crash surrogate measures when assessing crash risk.
> [13, Page ii]

Despite the limitations of defining and detecting "near-crash" events, the study identified roughly 30 times as many "near-crashes" as crash events [32, Page 141]. The study also identified, particularly in the many scatter plots of measured data, the serious problems that arise from computations of traffic characteristics (*e.g.*, TTC and IMU) that are not monotonic: more severe traffic hazards do not always result in a value that reflects a greater degree of hazard or risk than less severe hazards.

The majority of near-miss reporting systems rely on observer judgment as to whether a near-miss occurred or not, and at most have an informal qualitative assessment of severity. These approaches are not suitable for determination of traffic risk and safety because of the wide range of degree or severity of near-misses in traffic: everything from inconsequential movement at large distances and low relative speeds to inches of separation at high speed. Systems and approaches that count the occurrence of each near-miss event are not useful for the determination of the safety of vehicular traffic.

3 Existing Surrogate Measures

A number of existing non-collision approaches to determining traffic safety are summarized here.

3.1 Historical Collision Statistics

A common practice is to measure the safety of the behavior of a vehicle on the basis of the frequency of occurrence of collisions. Typical collision occurrence data is shown in Table 1.

Some drivers make less safe decisions (and take less safe actions) than others; however, the infrequency of collisions, and the many factors beyond driver decision-making that contribute to the occurrence of collisions, render historical collision statistics of limited use to evaluate driver performance, or to judge whether a driver (automated or human) is sufficiently safe to drive, particularly in congested and complex scenarios.

Table 1. Rates of involvement in all police-reported crashes, injury crashes, and fatal crashes per 100 million miles driven in relation to driver age, United States, 2014–2015. [33]

Age of Driver	All Crashes	Injury Crashes	Fatal Crashes
16–17	1,432	361	3.75
18–19	730	197	2.47
20–24	572	157	2.15
25–29	526	150	1.99
30–39	328	92	1.20
40–49	314	90	1.12
50–59	315	88	1.25
60–69	241	67	1.04
70–79	301	86	1.79
80+	432	131	3.85

3.2 Traffic Conflicts

The Traffic Conflicts Technique (TCT) was first introduced by Perkins and Harris at General Motors Research in 1967 [8], first published in 1968 [6], and codified into a procedures manual in 1969 [25] (Fig. 1).

> The basic types of accidents at intersections are left-turn, weave, cross-traffic, red-light violation, and rear-end incidents. For these five basic categories, over 20 objective conflict criteria have been defined for specific potential accident patterns at intersections.
> [6, Page 36]

> Within the four basic categories of rear-end incidents, over 10 specific rear-end conflicts have been defined.
> [6, Page 38]

> The traffic conflict technique delineates initial causes of potential accident situations. Over 10 specific categories of rear-end incidents have been defined.
> [6, Page 43]

Traffic conflicts enable an observer to categorize interactions among vehicles in traffic into a number of different conflict categories. However, other than counting the number of such conflicts by category, traffic conflicts provide no quantitative, nor objective, measure of collision hazard or risk.

> Campbell and King (1970) [34] used the General Motors TCT to measure the accident potential of two rural intersections. They found no significant association between conflicts and reported accidents.
> [35, Page 27]

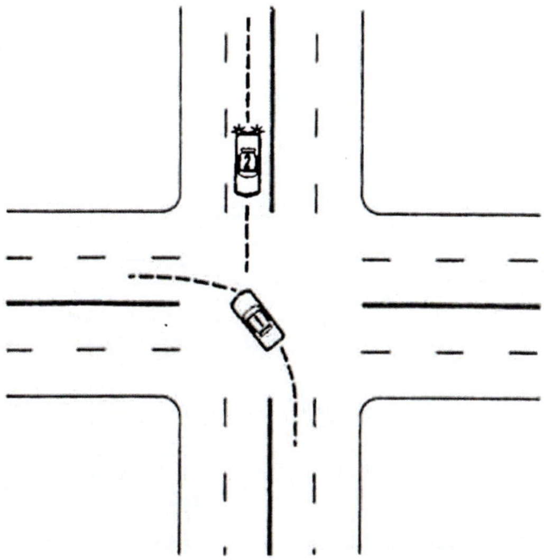

Fig. 1. Left-turn conflict. [6, Fig. 1, Page 36]

3.3 Time-to-Collision (TTC)

Some practitioners in the field of traffic safety use an estimate of time-to-collision or TTC to indicate whether the vehicle being analyzed is in a condition of high likelihood of an impending collision.

Hayward defined time to collision as follows:

> [T]he measure is the time required for two vehicles to collide if they continue at their present speeds and on the same path.
> [7, page 27]

An illustration of TTC is shown in Fig. 2 [35].

This measure has significant limitations, particularly in that the time until a collision will occur is highly dependent on the speed of the vehicle, the movements of the object with which it might collide, and the road conditions, none of which are incorporated into the time-to-collision measure [36].

Brown noted a particular type of problem with TTC:

> Each of the standard crash measures has weaknesses that restrict its utility. Minimum type I and type II TTC provide a continuous measure of how severe a situation resulted from the driver's response to the event so long as the driver does not collide with the other vehicle. For this measure, the larger the TTC, the safer the response. When the driver collides, however, the minimum TTC is zero regardless of whether the driver barely nudges

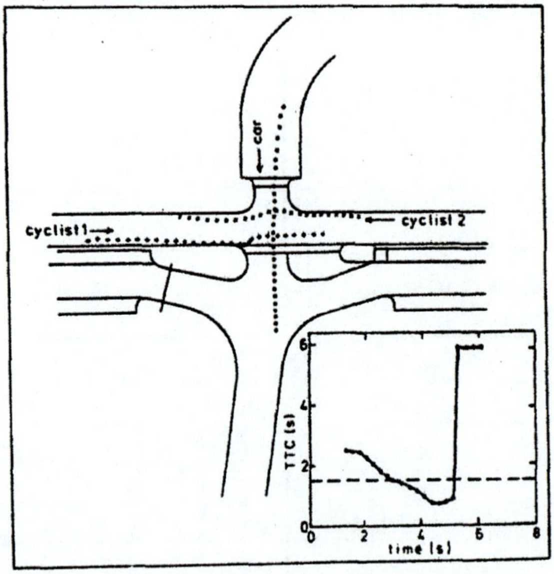

Fig. 2. Example of a serious conflict between a car from the minor road and cyclist 1. Bottom right: time-to-collision (TTC) curve. [35, Fig. 7, Page 32]

> the other vehicle with a small differential velocity or slams into the vehicle with a differential velocity of 70 mph.
> [36, page 42]

van der Horst identified one of the key limitations of TTC:

> [T]he relationship between TTC_{min} and conflict severity scores is not unambiguous; severe conflicts have a low TTC_{min}, but not all conflicts with a low TTC_{min} are regarded as severe.
> [37, page 107]

A higher-hazard encounter, *e.g.*, one that has the potential to result in a collision with high relative speed (and therefore would produce significant damage and/or injury), can have a larger (and therefore apparently less concerning) TTC than a lower-hazard encounter, such as one where the potential collision would occur with nearly zero relative speed. This example illustrates that TTC is not monotonic with hazard and risk, as will be discussed further below.

Many attempts to improve TTC have been made, principally focusing on assumed behavior (*e.g.*, deceleration) of the following vehicle.

> Time to collision is an important time-based safety indicator for detecting rear-end conflicts in traffic safety evaluations. A major weakness of the time to collision notion is the assumption of constant velocities during the course of an accident.
> ...
> Results indicate that in the third case (linear acceleration), the average duration of exposure to critical time to collision values is greater than the others. So, applying time to collision based on the assumption of linear acceleration in collision avoidance systems would decrease driver errors more than other cases.
> [38, page 294]

These limitations render TTC unsuitable as a general measure of collision risk in traffic.

3.4 Post-Encroachment Time (PET)

The concept of Post-Encroachment Time was introduced by Allen [4] in 1978:

> Post encroachment time (PET) for a conflict is identified as the time from the end of encroachment to the time that the through vehicle actually arrives at the potential point of collision (T_4-T_2 in Figure 3). This is an obvious measurement of how nearly a collision has been avoided. PET is also a suitable measurement for identifying the resulting events in the final stage of a traffic conflict. Although it directly describes neither the situation defined in the initial stage nor the action taken by the drivers in the intermediate stage, it does represent the result of the combined effects of the two earlier stages. For example, a PET value approaching

zero demonstrates that a collision was avoided by only the very smallest of margins. This could result from a very severe situation perceived in the initial stage, a very poor driving maneuver during the intermediate stage, or a combination of the two.
[4, Page 70]

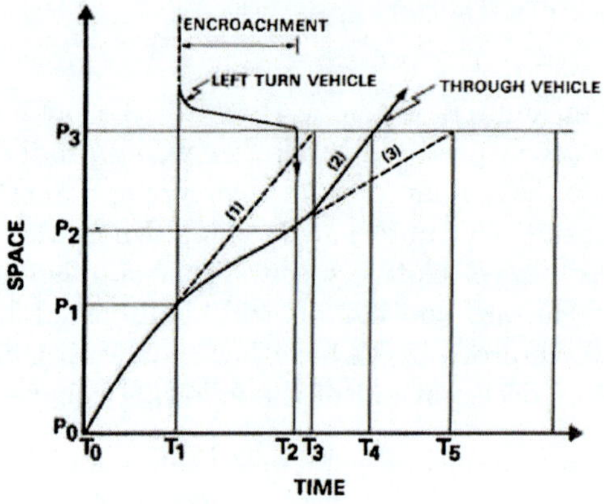

Fig. 3. Time-space diagram of a typical left-turn conflict. [4, Fig. 4, Page 70]

The PET value, ... is a measure that also includes the 'near misses'. It is defined as the time between the moment that the first road-user leaves the path of the second [t_1] and the moment that the second reaches the path of the first [t_2] (see Fig. 4). The PET value indicates the extent to which they missed each other. In urban areas, PET values of one second and lower are indicated as possibly critical.
[39, Page 361]:

Each of the preceding conflict measures [including PET] incorporates a degree of weakness by its very definition. However, it was felt that each had the potential to more adequately explain collision occurrence than the conventional brake application procedure. In particular, one would expect that those measurements that identified events in or near the final stage of the conflict-generation sequences would possess the greatest explanatory power.
[4, Page 70]

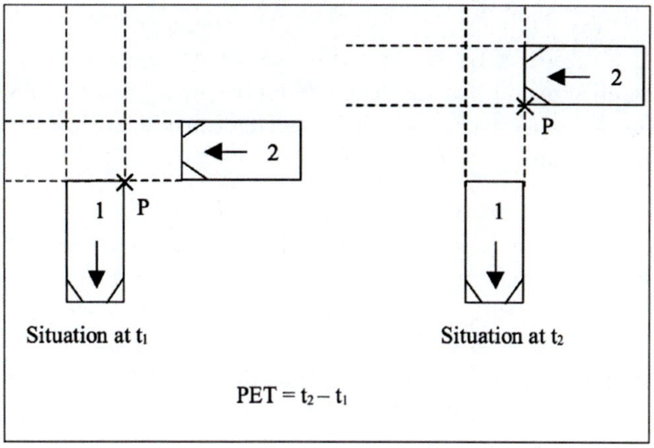

Fig. 4. Definition of post-encroachment time (PET). [39, Fig. 6, Page 362]

> The authors' recommendation that PET is an appropriate conflict technique capable of explaining more about collision occurrence than brake applications is also questionable. The recommendation is based on the analysis of only the left-turn maneuver and was not applied to include other major conflict types commonly found at intersections.
> Martin R. Parker, Jr., Virginia Highway and Transportation Research Council [4, Page 73]

PET was developed specifically for traffic conflicts arising from unprotected left turns, where a vehicle in an intersection crosses the path of an oncoming vehicle. While attempts have been made to modify PET to apply to other traffic situations, PET is not generally applicable.

While PET is a surrogate measure based on non-collision events, it is a *retrospective* measure, looking back in time to determine by what interval of time a collision was avoided. These are the principal weaknesses of the measure, as noted by Allen [4, Page 70] and Parker [4, Page 73].

3.5 Responsibility-Sensitive Safety (RSS)

Responsibility-Sensitive Safety (RSS) is a set of five rules intended to ensure that automated vehicles operate safely [5]. These are all sensible rules; however, the full complexity of driving simply cannot be captured in five rules.[1]

01. Safe Distance Enforce a safe following distance from a vehicle ahead, based on vehicle speed and stopping ability.
02. Cutting In Merge into a lane with sufficient lateral distance from other vehicles.

[1] https://www.mobileye.com/responsibility-sensitive-safety/.

03. Right of Way Give right of way to other vehicles.
04. Limited Visibility Be cautious in areas of limited visibility.
05. Avoid Collisions [I]f an object suddenly appears in the AV's direct path, the AV must avert a crash by veering into the next lane, provided it would not cause a different collision.

While the first two rules include a quantitative measurement of d_{min}, which is defined as the minimum safe distance for those two maneuvers, RSS does not provide any sort of quantitative measurement of vehicle safety in traffic. The first two are behavioral rules that are to be observed, rather than a quantitative measure of how safely a vehicle is behaving. The remaining three rules are important, but informal and not quantitative, and provide no utility in measuring the safety of the operation of a vehicle. Instead, RSS is a set of guidelines for automated vehicle control decision-making.

Methods that use rules for particular maneuvers or situations, such as RSS, will not scale well to even a portion of the full range of traffic scenarios encountered in real traffic, because of the large number of rules required to accommodate each different type of maneuver.

Rules 3, 4, and 5 assume that vehicle controllers will behave accordingly, and will have the capability to do so.

A continuous quantitative measure of collision hazards in traffic, such as the novel surrogate measure reviewed below, can assess the risk and safety of a vehicle being operated in accordance with the RSS rules.

3.6 Instantaneous Safety Metric (ISM)

Instantaneous Safety Metric (ISM), developed by the U.S. National Highway Traffic Safety Administration [2], predicts all future positions of one or more vehicles, and examines the overlap of future reachable regions to determine whether there is a "critical" or "imminent" overlap of regions. A flowchart of the ISM assessment is shown in Fig. 5. An extended approach (Model Predictive Instantaneous Safety Metric or MPrISM) is presented in [3].

To determine whether a future interaction is "critical" or "imminent" requires the determination of "the probability of the driver choosing to pursue a set of accelerations" [2, page 6], in other words, a prediction of the actions of the operator of each vehicle is required.

Importantly, the result of computing the future reachable regions and the possible regions of overlap is one of the four outcomes listed below.

> There are four possible combinations resulting from interaction between the possible and unavoidable spaces of two vehicles (Vehicles A & B in this case). [40, page 20]

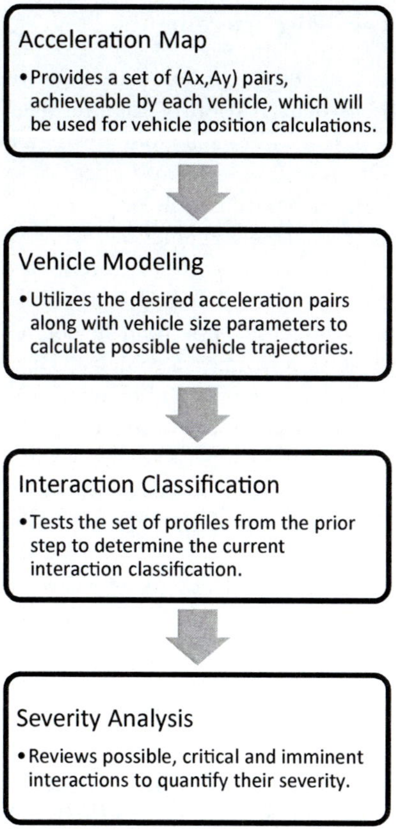

Fig. 5. ISM process flow chart. [2, Fig. 3, Page 4]

1. The possible space of both vehicles overlap. (Possible Interaction)
2. The unavoidable spaces of both vehicles overlap. (Imminent Interaction)
3. The unavoidable space of Vehicle A overlaps the possible space of Vehicle B. (Critical interaction for Vehicle A)
4. The possible space of Vehicle A overlaps the unavoidable space of Vehicle B. (Critical Interaction for Vehicle B)

This is not a quantitative measure of risk of collision; it is a categorical indicator rather than a metric. The ISM can determine whether it is possible for an "interaction" to occur in the future, but other than identifying the future interaction as being either Possible, Imminent, or Critical, the ISM provides no quantitative indication of the degree of risk nor the severity of the potential consequences.

3.7 AV Control System Disengagements

The number of disengagements of the onboard decision-making system per mile (where a disengagement is a deactivation or manual override of the automated system, such as described in the California Code of Regulations, Title 13, Div. 1, Ch. 1, Article 3.7, §227.50[2]) is also used as an indication of the performance and safety of the behavior of an automated vehicle.

The key assumption is that "disengagements" occur when hazardous driving conditions are encountered that the controller cannot or does not handle safely, and control of the vehicle must be transferred to the onboard safety driver.

The number of "disengagements" per mile is not a useful measure of the behavior or safety or risk of an automated vehicle. "Disengagements" can have many causes which may not be related to the behavior or decision-making system of the vehicle, they are not repeatable, are subject to the judgment of the safety driver and therefore occur due to subjective considerations and as a result are not objective, and are influenced by the selection of the conditions and scenarios under which the vehicle is operated and the operational policy or policies under which the driver operates [42].

As a result, disengagement rate is not an effective measure of the hazards encountered by an automated vehicle, and is at best an indirect indicator of operational risk and safety.

3.8 Inertial Measurement Unit (IMU)

A number of current approaches to measuring risk and safety of vehicles utilize data from an onboard inertial measurement unit (IMU). These data can indicate rapid deceleration ("hard braking") or rapidly executed turns ("swerving"), and it is thought that events of this type, with accelerations above a threshold, reflect unsafe operation of the subject vehicle.

While IMU data are readily available, either from onboard accelerometers or from onboard electronics such as mobile phones, smooth driving (*i.e.*, with consistently low levels of acceleration) is at best a poor proxy for the risk or safety of the operation of a vehicle, for two important reasons:

[2] 13 CCR §227.50 Reporting Disengagement of Autonomous Mode:

> For the purposes of this section, "disengagement" means a deactivation of the autonomous mode when a failure of the autonomous technology is detected or when the safe operation of the vehicle requires that the autonomous vehicle test driver disengage the autonomous mode and take immediate manual control of the vehicle, or in the case of driverless vehicles, when the safety of the vehicle, the occupants of the vehicle, or the public requires that the autonomous technology be deactivated. (b) Every manufacturer authorized under this article to test autonomous vehicles on public roads shall prepare and submit to the department an annual report summarizing the information compiled pursuant to subsection (a) by January 1st, of each year. [41]

1. IMU data are blind to other traffic objects, and therefore necessarily take no account of how the subject vehicle is moving in relation to these other traffic objects. As a result, IMU data reflect nothing about how the subject vehicle is interacting with traffic.
2. Many cases have demonstrated events where a vehicle was smoothly driven into a collision. Similarly, hard braking occurs when a skilled driver avoids a collision, for example in the classic example of a child chasing a ball into a lane of traffic from between two parked cars.

The use of IMU data relies on the assumption that rapid decelerations are related to risky driving behaviors. The lack of correlation of IMU data with risk and safety of the operation of a vehicle makes it unsuitable for use as a measure.

4 Novel Surrogate Measure

4.1 Collision Hazard Measure (SHM)

A novel measure of vehicle and traffic risk and safety (SHM) introduced in 2022 [43] uses the position and velocity of the subject vehicle, the position and velocity of each traffic object, the road conditions, and an estimate of the maneuverability of the subject vehicle and traffic objects (maximum safe braking deceleration rate and maximum safe turning rate).

In all cases, the measure is computed sequentially for the subject vehicle in relation to each traffic object.

The measure incorporates the square of the relative speed between the subject vehicle and a traffic object (S_{rel}) divided by the distance that separates the vehicle and the object (d_{sep}).

$$m = S_{rel}^2 / d_{sep} \tag{1}$$

m has the units of $\left[\text{length}/\text{time}^2\right]$ or [acceleration].

This measure has the essential character of near-misses described above, combining proximity (separation distance) and motion (relative speed) for each pair of traffic objects in a traffic scenario (*e.g.*, car_i-pedestrian_j) at each time-step or frame of sensor data. In this case, relative speed is in the numerator, so the measure will be larger for larger values of relative speed; separation distance is in the denominator so that the measure will be larger for *smaller* separation distances.

This matches our perception of near-misses. A vehicle that is moving at 0.5 m/s (1 mph) past a pedestrian at a distance of 1 m (39 in.) would not be alarming or considered to be particularly dangerous. In contrast, a vehicle that is moving at 30 m/s (67 mph) past a pedestrian at the same distance would be highly alarming and would be considered to be seriously dangerous. Both relative speed and separation distance are essential characteristics of a quantitative measure of near-misses, and the hazard that they produce.

The influence of the speed of the subject vehicle in relation to the traffic object is considerably magnified compared to other approaches to quantify near-misses. This magnification is desired, and is an important characteristic of the measure since the square of the speed is directly proportional to the kinetic energy of the subject vehicle in relation to the traffic object, and the dissipation of kinetic energy in a collision is the cause of damage and injury.

This approach to combining a compensated (*i.e.*, squared) value of relative speed with separation distance has the essential characteristic of monotonicity: a less severe traffic hazard will result in a lower numerical value of the measure than a more severe traffic hazard.

Importantly, the determination of the measure values makes no assumptions nor predictions about the behavior or actions of traffic objects; it simply assesses the current state of the near-miss interaction between each pair of objects, and does not predict future actions, decisions, or trajectories.

5 Characteristics

The key characteristics of any effective collision hazard measure include:

1. **Leading:** The measure determines hazards without the occurrence of collisions, in contrast to *trailing* measures that determine hazards after a number of collisions have occurred.
2. **Quantitative:** The result of the computation of the measure is a numerical representation of the collision hazard encountered.
3. **Continuous:** The quantitative result is on a continuous scale, *e.g.*, from 0 (safe) to 100 (a few centimeters from collision), and is nearly continuous in time, subject only to the update rate of the traffic sensor(s) used.
4. **Independent:** The computation of the measure relies only on external observation of vehicles and traffic objects, and road/street surface conditions, and not on sensor data or decision-making involved in the control of the vehicle. The measure considers each traffic object to be a "black box", not subject to internal scrutiny.
5. **Direct:** The measure directly determines the hazard between traffic objects, rather than indirect or proxies for hazard.
6. **Repeatable:** Observation of the same behavior in the same traffic scenario will produce the same quantitative value of the collision hazard measure.
7. **No Assumptions:** The measure does not make use of assumptions or predictions about the actions or behaviors of traffic objects.
8. **Monotonic:** A more severe collision hazard will always result in a larger value of the collision hazard measure. Two identical collision hazards, generated by different traffic conditions, will always result in the same value of the collision hazard measure. To be at all useful, the measure should at least conform with the properties of an *Ordinal Scale*, preferably a *Ratio Scale* [44,45].
9. **Objective:** No qualitative or subjective input is included in the computation of the result. The result depends solely on the measured kinematics of the vehicles and traffic objects, and road/street surface conditions.

10. **Computable:** Given the kinematics (positions and velocities) of the subject vehicle and other traffic objects in a scenario, and estimates of the capabilities of the subject vehicle to stop and turn, the measure can be automatically calculated by a machine such as a computer.
11. **Scalable:** The computation of the measure does not depend on the complexity or other characteristics of the traffic scenario, and therefore naturally and easily scales up to the full range of situations and scenarios encountered in real traffic. Measures that comprise individual rules or computations for specific traffic scenarios are inherently not scalable.

5.1 Comparison of Surrogate Collision Hazard Measures

Table 2 presents a comparison of the characteristics of the traffic collision hazard measures briefly reviewed here. Note that only the novel SHM measure [43] satisfies all of the key characteristics required for any effective surrogate collision hazard measure.

Table 2. Characteristics of Selected Traffic Collision Hazard Measures

Characteristic		Traffic Conflicts	TTC	PET	RSS	ISM	AV Disengagements	IMU	SHM
1.	Leading		✓	[a]	✓	✓			✓
2.	Quantitative		✓	✓	binary	binary	binary	✓	✓
3.	Continuous	count[b]	✓		count[b]	count[b]	count[b]	count[b]	✓
4.	Independent		✓	✓	✓	✓		✓	✓
5.	Direct		✓	✓	✓	✓			✓
6.	Repeatable		✓	✓	✓	✓		✓	✓
7.	No Assumptions			✓					✓
8.	Monotonic								✓
9.	Objective		✓	✓		✓			✓
10.	Computable		✓	✓	partial[c]	✓		✓	✓
11.	Scalable		✓		✓			✓	✓

[a] PET is non-collision but *retrospective* and therefore is not a leading measure.
[b] Occurrences of events are counted, and therefore are not a continuous measure.
[c] Two of the RSS rules are computable; the remainder are not.

6 Conclusions

Surrogate measures of traffic safety, based on non-collision interactions and events, are an important improvement on the use of collision statistics as a means of assessing the safety of roadways, intersections, vehicles, or mobility systems.

Leading indicators for measuring and assessing the safe operation of vehicles in traffic are essential to the deployment of automated mobility. Existing approaches, such as Traffic Conflicts, TTC, PET, RSS, ISM, IMU data, and AV control system disengagements do not provide the information required to provide a measurement of safety that will be useful to regulators and the public.

The existing methods listed above require assumptions and/or predictions about the behaviors of traffic objects, which will not, in general, be correct, limiting the value of the information that they provide. Several methods, such as TTC and PET, were developed for particular traffic interactions (following traffic and left turns, respectively), and do not apply well to general traffic interactions.

Lagging indicators, such as historical collision statistics, do not provide timely information, and require the ethically unacceptable occurrence of collisions, property damage, injuries, and deaths.

Table 2 lists the essential characteristics to provide effective and consistent surrogate measures of traffic risk and safety that will be useful to ADS stakeholders, including AV developers, traffic infrastructure developers and managers, regulators, legislators, and the public.

The comparison shows that the novel collision hazard measure (SHM) overcomes the limitations of existing measures, provides an independent leading indication of safety, and does not require assumptions nor predictions of behaviors.

References

1. Browne, D.: Automated Driving Systems (ADS) - An introduction to technology and vehicle connectivity - Part 3, Association for the Advancement of Automotive Medicine (AAAM). https://www.aaam.org/automated-driving-systems-ads-introduction-technology-vehicle-connectivity/. Accessed 4 Feb 2023
2. Every, J.L., Barickman, F., Martin, J., Rao, S., Schnelle, S., Weng, B.: A novel method to evaluate the safety of highly automated vehicles. In: 25th International Technical Conference on the Enhanced Safety of Vehicles (ESV) National Highway Traffic Safety Administration, Detroit, Michigan (2017). https://www-esv.nhtsa.dot.gov/Proceedings/25/25ESV-000076.pdf
3. Weng, B., Rao, S.J., Deosthale, E., Schnelle, S., Barickman, F.: Model predictive instantaneous safety metric for evaluation of automated driving systems. In: 2020 IEEE Intelligent Vehicles Symposium (IV), pp. 1899–1906. IEEE (2020). https://ieeexplore.ieee.org/abstract/document/9304635
4. Allen, B.L., Tom, S.B.: Analysis of traffic conflicts and collisions. Transp. Res. Rec. **667**, pp. 67–74 (1978). Paper sponsored by Committee on Methodology for Evaluating Highway Improvements, Transportation Research Board, U.S. National Academies of Science, Engineering, Medicine. https://onlinepubs.trb.org/Onlinepubs/trr/1978/667/667-009.pdf
5. Shalev-Shwartz, S., Shammah, S., Shashua, A.: On a formal model of safe and scalable self-driving cars. arXiv preprint arXiv:1708.06374 (2017). https://arxiv.org/pdf/1708.06374.pdf
6. Perkins, S.R., Harris, J.I.: Traffic conflict characteristics: accident potential at intersections. Highway Res. Rec. **225**, 35–43 (1968). Paper sponsored by Committee on Traffic Safety and presented at the 47th Annual Meeting. https://onlinepubs.trb.org/Onlinepubs/hrr/1968/225/225-004.pdf
7. Hayward, J.C.: Near-miss determination through use of a scale of danger. In: Proceedings of the 51st Annual Meeting of the Highway Research Board. Washington, DC: Transportation Research Board, U.S. National Academies of Science, Engineering, Medicine, January 1972, pp. 24–34. https://onlinepubs.trb.org/Onlinepubs/hrr/1972/384/384-004.pdf

8. Perkins, S.R., Harris, J.I.: Traffic conflict characteristics: accident potential at intersections. General Motors Research Publication, Lansing, MI, Technical report GMR-718, December 1967
9. Güttinger, V.A.: From accidents to conflicts: alternative safety measurement. In: Proceedings of the 3rd International Workshop on Traffic Conflicts Techniques, Kraay, J.H. (ed.) no. R-82-27. Leidschendam, The Netherlands: The International Committee on Traffic Conflicts Techniques ICTCT, April 1982, pp. 14–25 (1982). https://www.ictct.net/wp-content/uploads/XX-Leidschendam-1982/1982-Proceedings.pd#page=14
10. Gettman, D., Head, L.: Surrogate safety measures from traffic simulation models, final report," U.S. Department of Transportation, Federal Highway Administration, Office of Safety Research and Development, Technical report, FHWA-RD-03-050, January 2003. https://www.fhwa.dot.gov/publications/research/safety/03050/
11. Gettman, D., Pu, L., Sayed, T., Head, L.: Surrogate safety assessment model and validation: Final report, U.S. Department of Transportation, Federal Highway Administration, Office of Safety Research and Development, Tech. Rep. FHWA-HRT-08-051, June 2008. https://www.fhwa.dot.gov/publications/research/safety/08051/
12. Tarko, A., Davis, A.G., Saunier, N., Sayed, T., Washington, S.: Surrogate measures of safety, ANB20(3) Subcommittee on Surrogate Measures of Safety, ANB20 Committee on Safety Data Evaluation and Analysis, Transportation Research Board, U.S. National Academies of Science, Engineering, Medicine, pp. 1–13, January 2009, white Paper. https://www.researchgate.net/publication/245584894_Surrogate_Measures_of_Safety
13. Guo, F., Klauer, S.G., McGill, M.T., Dingus, T.A.: Evaluating the relationship between near-crashes and crashes: can near-crashes serve as a surrogate safety metric for crashes? United States. Department of Transportation. National Highway Traffic Safety Administration, Technical report DOT-HS-811-382, October 2010. https://www.nhtsa.gov/sites/nhtsa.gov/files/documents/dot_hs_811_382.pdf
14. Ariza, A.: Validation of road safety surrogate measures as a predictor of crash frequency rates on a large-scale microsimulation, Master's thesis, University of Toronto, Department of Civil Engineering, Toronto, Canada, June 2011. https://tspace.library.utoronto.ca/bitstream/1807/30160/1/Ariza_Alexander_201111_MASc_thesis.pdf
15. Yang, H.: Simulation-based evaluation of traffic safety performance using surrogate safety measures. Ph.D. dissertation, Rutgers University, Department of Civil and Environmental Engineering, New Brunswick, NJ, January 2012. https://rucore.libraries.rutgers.edu/rutgers-lib/36680/
16. Lorion, A.C., Persaud, B.: Investigation of surrogate measures for safety assessment of urban two-way stop controlled intersections. Can. J. Civ. Eng. **52**(12), 987–992 (2015). https://cdnsciencepub.com/doi/pdf/10.1139/cjce-2015-0023?download=true
17. Ušpalytė-Vitkūnienė, R., Laureshyn, A.: Perspectives for surrogate safety studies in East-European countries. Baltic J. Road Bridge Eng. **3**, 161–166 (2017). https://bjrbe-journals.rtu.lv/article/download/bjrbe.2017.19/1762
18. Johnsson, C., Laureshyn, A., De Ceunynck, T.: In search of surrogate safety indicators for vulnerable road users: a review of surrogate safety indicators. Transp. Rev. (2018). https://ictct.net/wp-content/uploads/SMoS_Library/LIB_Johnsson_2018.pdf

19. Tarko, A.: Measuring Road Safety with Surrogate Events. Elsevier Inc, Amsterdam (2019). https://www.elsevier.com/books/measuring-road-safety-using-surrogate-events/9780128105047
20. Johnsson, C.: Surrogate measures of safety with a focus on vulnerable road users: an exploration of theory, practice, exposure, and validity, Ph.D. dissertation, Lund University, Faculty of Engineering, Department of Technology and Society, Transport and Roads, Lund, Sweden, September 2020. https://portal.research.lu.se/en/publications/surrogate-measures-of-safety-with-a-focus-on-vulnerable-road-user
21. Saunier, N., Laureshyn, A.: Surrogate measures of safety. In: Vickerman, R. (ed.) International Encyclopedia of Transportation, pp. 662–667. Elsevier, Oxford (2021). https://www.sciencedirect.com/science/article/pii/B9780081026717101976
22. Ambros, J., Jurewicz, C., Chevalier, A., Valentová, V.: Speed-related surrogate measures of road safety based on floating car data. In: Macioszek, E., Sierpiński, G. (eds.) Research Methods in Modern Urban Transportation Systems and Networks. LNNS, vol. 207, pp. 129–144. Springer, Cham (2021). https://doi.org/10.1007/978-3-030-71708-7_9
23. Lu, J., Grembek, O., Hansen, M.: Connecting surrogate safety measures to crash probablity via causal probabilistic time series prediction, October 2022. https://arxiv.org/abs/2210.01363
24. Sayed, T., Tarko, A., Trivedi, M. (eds.): Accident Analysis & Prevention, vol. 148–161, 2020–2021, special Issue: Crash Precursors. https://www.sciencedirect.com/journal/accident-analysis-and-prevention/special-issue/101VCPS49HX
25. Perkins, S.R.: Traffic conflicts technique procedures manual. General Motors Research Publication, Warren, MI, Technical report GMR-896, August 1969
26. Wishart, J., et al.: Driving safety performance assessment metrics for ADS-equipped vehicles. SAE Technical Paper, vol. 2, no. 2020-01-1206 (2020). https://www.researchgate.net/publication/340652968_Driving_Safety_Performance_Assessment_Metrics_for_ADS-Equipped_Vehicles
27. Heinrich, H.W.: Industrial Accident Prevention. McGraw-Hill Book Company Inc, A Scientific Approach. New York & London (1931)
28. Busch, C.: Preventing Industrial Accidents: Reappraising HW Heinrich-More than Triangles and Dominoes. Routledge (2021). https://www.routledge.com/Preventing-Industrial-Accidents-Reappraising-H-W-Heinrich-More-than/Busch/p/book/9780367343804
29. Van der Schaaf, T.W., Lucas, D.A., Hale, A.R.: Near miss reporting as a safety tool. Butterworth-Heinemann. Elsevier Ltd., Oxford (1991). https://www.elsevier.com/books/near-miss-reporting-as-a-safety-tool/van-der-schaaf/978-0-7506-1178-7
30. Aviation safety reporting system. Wikipedia, Last updated: 19-FEb-2022. https://en.wikipedia.org/wiki/Aviation_Safety_Reporting_System. Accessed 10 Apr 2022
31. Phimister, J.R., Oktem, U., Kleindorfer, P.R., Kunreuther, H.: Near-miss incident management in the chemical process industry. Risk Anal. Int. J. **23**(3), 445–459 (2003). https://riskcenter.wharton.upenn.edu/wp-content/uploads/2014/07/03-01-JP.pdf
32. Dingus, T.A., et al.: The 100-car naturalistic driving study, Phase II-results of the 100-car field experiment," United States. Department of Transportation. National Highway Traffic Safety Administration, Technical report DOT-HS-810-593, April 2006. https://rosap.ntl.bts.gov/view/dot/37370/dot_37370_DS1.pdf

33. Rates of motor vehicle crashes, injuries and deaths in relation to driver age, united states, 2014–2015.AAA Foundation for Traffic Safety, June 2017. https://aaafoundation.org/rates-motor-vehicle-crashes-injuries-deaths-relation-driver-age-united-states-2014-2015/
34. Campbell, R.E., King, L.E.: The traffic conflicts technique applied to rural intersections. Accid. Anal. Prevent. **2**(3), 209–221 (1970). https://www.sciencedirect.com/science/article/abs/pii/0001457570900436
35. van der Horst, R.A.: The analysis of traffic behaviour by video. In: Kraay, J.H. (ed.) Proceedings of the 3rd International Workshop on Traffic Conflicts Techniques, no. R-82-27. Leidschendam, The Netherlands: The International Committee on Traffic Conflicts Techniques ICTCT, April 1982, pp. 26–41. https://www.ictct.net/wp-content/uploads/XX-Leidschendam-1982/1982-Proceedings.pdf#page=26
36. Brown, T.L.: Adjusted minimum time-to-collision (TTC): a robust approach to evaluating crash scenarios. In: Proceedings of the Driving Simulation Conference North America, vol. 40, 2005, pp. 40–48. https://www.nads-sc.uiowa.edu/dscna/2005/papers/Adjusted_Minimum_TimeToCollision.pdf
37. van der Horst, R.A.: Video analysis of road user behaviour at intersections. In: Van der Schaaf, T.W., Lucas, D.A., Hale, A.R. (eds.) Near Miss Reporting as a Safety Tool, pp. 93–109. Butterworth-Heinemann, Oxford (2013), ch. 9. https://www.elsevier.com/books/near-miss-reporting-as-a-safety-tool/van-der-schaaf/978-0-7506-1178-7
38. Saffarzadeh, M., Nadimi, N., Naseralavi, S., Mamdoohi, A.R.: A general formulation for time-to-collision safety indicator. In: Proceedings of the Institution of Civil Engineers-Transport, vol. 166, no. 5. Thomas Telford Ltd, 2013, pp. 294–304. https://www.icevirtuallibrary.com/doi/abs/10.1680/tran.11.00031
39. van der Horst, R.A., de Goede, M., Hair-Buijssen, S., Methorst, R.: Traffic conflicts on bicycle paths: a systematic observation of behaviour from video. Accid. Anal. Prevent. **62**, 358–368 (2014). https://www.sciencedirect.com/science/article/abs/pii/S0001457513001401
40. Every, J.L., Martin, J., Barickman, F., Rao, S., Schnelle, S., Weng, B.: A method for evaluating automated vehicle safety. SAE Government Industry Meeting (2017). https://www.nhtsa.gov/sites/nhtsa.gov/files/documents/sae2017jevery.pdf
41. Reporting disengagement of autonomous mode. California Code of Regulations, Title 13. Motor Vehicles, Division 1. Department of Motor Vehicles, Chapter 1. Department of Motor Vehicles, Article 3.7. Testing of Autonomous Vehicles, 2021, 13 CCR §227.50, current through 2/17/23 Register 2023, No. 7. https://govt.westlaw.com/calregs/Document/I5C2092835A1E11EC8227000D3A7C4BC3
42. Khattak, Z.H., Fontaine, M.D., Smith, B.L.: Exploratory investigation of disengagements and crashes in autonomous vehicles under mixed traffic: an endogenous switching regime framework. IEEE Trans. Intell. Transp. Syst. **22**(12), 7485–7495 (2020). https://www.osti.gov/servlets/purl/1649156
43. Antonsson, E.K.: A general measure of collision hazard in traffic (2022). https://arxiv.org/abs/2205.08640
44. Stevens, S.S.: On the theory of scales of measurement. Science **103**(2684), 677–680 (1946). http://www.jstor.org/stable/1671815
45. Measurement scale. Encyclopædia Britannica. https://www.britannica.com/topic/measurement-scale. Accessed 10 Apr 2022

46. IEEE VT/ITS/AV Decision Making Working Group. White paper-literature review on kinematic properties of road users for use on safety-related models for automated driving systems. IEEE, Technical report (2022). https://ieeexplore.ieee.org/stamp/stamp.jsp?arnumber=9763462
47. Intelligent Transportation Systems Committee. IEEE standard for assumptions in safety-related models for automated driving systems. IEEE Std 2846–2022 (2022). https://standards.ieee.org/ieee/2846/10831/

Introducing ODD-SAF: An Operational Design Domain Safety Assurance Framework for Automated Driving Systems

Matteo Oldoni[1](✉) and Siddartha Khastgir[2]

[1] NVIDIA Corporation, Milan, Italy
moldoni@nvidia.com
[2] WMG University of Warwick, Coventry, UK
s.khastgir.1@warwick.ac.uk

Abstract. Over the past years, there has been an increasing acceptance on the need for scenario-based testing to ensure safe performance of Automated Driving Systems (ADS). This is a departure from earlier conceptions where number of miles driven was being considered as the only way of demonstrating safety. As ADS show a great degree of variety in their complexity, use cases, as well as Operational Design Domain (ODD), a scalable and pragmatic approach for safety assurance of ADS, the ODD-SAF, is therefore proposed. The ODD-SAF relies on the ODD description and leverages on the EU and UNECE discussions around ADS safety requirements and assessment methods to generate behavioural competencies for the overall safety assurance. The approach extends with the identification of test scenarios, classified into nominal, critical and failure types, and pass-fail criteria, leveraging for example the concept of rules of the road and safety models for driving behaviour. It is suggested that a SAF should incorporate each of the categories of the test scenarios to ensure confidence in the performance of the ADS.

Keywords: ODD · AV Safety Assurance Framework · Scenario Generation · Behavioural Competencies · Automated Driving Systems · AV Certification

1 Introduction

Due to increasing system complexity, certification of an Automated Driving System (ADS) poses challenges that cannot be addressed exclusively with the application of prescriptive requirements or assessment through a series of repeatable tests, as commonly found with traditional certification regimes.

Recent developments in the automotive industry and academia suggest the use of a scenario-based approach promoted by the system's Operational Design Domain (ODD) for safety assurance. In such a framework, the ODD defines the safe operating boundary and applicable behavioural competencies, whereas the scenarios set out individual test conditions.

G. Meyer and S. Beiker (Eds.): ARTSymposium 2022, LNMOB, pp. 133–151, 2023.
https://doi.org/10.1007/978-3-031-34757-3_11

This chapter describes how ODDs, scenarios, and behavioural competencies are bound by a scalable ODD safety assurance framework. The workflow utilises the system's ODD, the behavioural competencies derived from regulatory guidance and the rules of the road as input in order to map a set of representative testing scenarios. Scenarios are either generated to address unique behaviours, especially in critical and failure situations, or mapped from existing libraries, where labels help to identify applicable rules-related scenarios.

To complete the picture, a holistic multi-pillar assessment leveraging audit, simulations, and physical testing is being considered by regulators to ensure satisfactory proof of performance against the relevant scenarios.

1.1 The Importance of Appropriate ODD Taxonomy

According to a widely used definition, ODD refers to *"Operating conditions under which a given driving automation system or feature thereof is specifically designed to function, including, but not limited to, environmental, geographical, and time-of-day restrictions, and/or the requisite presence or absence of certain traffic or roadway characteristics."* (SAE J3016).

For a given ODD, it is crucial for the ADS to ensure:

- it can operate safely within its ODD
- it will be primarily used within its ODD
- it can monitor whether it is inside/outside its ODD, and consequently react to it

The conditions constituting the ODD in which the ADS is designed to operate play a key role to determine which ADS competencies are required. For example, if an ADS has an ODD, which comprises roads with non-signalised junctions, one of the required behaviour competencies for the ADS in that ODD could potentially be "unprotected left or right turn". However, the same behaviour competency may not be required if the ODD of an ADS is limited to motorways or highways with signalised junctions.

ODD description completeness is ensured by using an appropriate taxonomy. The BSI PAS 1883 [1] introduced a taxonomy, which contains a standardised set of ODD attributes covering scenery, environmental conditions, and dynamic elements. The SAE AVSC00002202004 [2] also presented a conceptual framework and lexicon for defining ODDs. The lexicon includes key attributes covering environmental conditions, road surface conditions, roadway infrastructure, operational constraints, road users, roadside objects and connectivity. On-going efforts are also seen in the ASAM OpenODD [3] project and the ISO 34503 [4] standard on ODD format and taxonomy. The ASAM OpenLABEL standard [5] also utilises an ODD and behaviour-based model for scenario tagging and organisation. Based on the review of ODD-related standards, there is consensus across the industry forming to address the question on what an ODD is and what attributes an ODD should contain.

As the ODD defines the operating conditions of the ADS, to be able to claim completeness, it also needs to underpin the scenario-generation process for testing the ADS – including considerations on coverage and safety metric(s), as shown in Fig. 1.

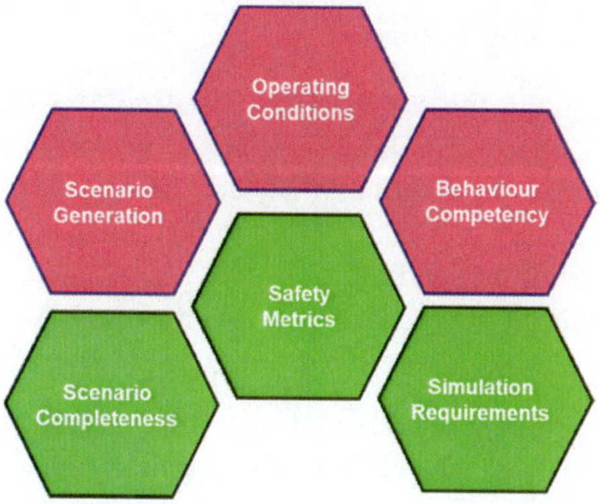

Fig. 1. Correlation between the ODD taxonomy and other pillars

1.2 Behaviour Competencies

The concept of "behavioural competencies" is useful in determining the safety of the performance of the Dynamic Driving Task (DDT) by an ADS. The Automated Vehicle Safety Consortium, or AVSC, has provided these definitions [7]:

- *Behaviour: Specific goal-oriented actions directed by an engaged ADS in the process of completing the DDT or DDT fallback within the ODD (if applicable) at a variety of timescales.*
- *Behavioural Competency: Expected and measurable capability of an ADS feature operating a vehicle within its ODD.*

Behavioural competencies can be described with different abstraction levels, similarly to functional, logical, and concrete scenarios. Refinement of the competencies from a functional to a more concrete level is possible by following the approach herein proposed. Such competencies track the three broad categories of driving situations that may be encountered in performance of the DDT: nominal, critical, and failure.

Nominal driving situations are those in which behaviour of other road users and the operating conditions of the given ODD are reasonably foreseeable (e.g., other traffic participants operating in line with traffic regulations) and no failures occur that are relevant to the ADS's performance of the DDT.

Critical driving situations are those in which the behaviour of one or more road users (e.g., violating traffic regulations, etc.) and/or a sudden and not reasonably foreseeable change of the operating conditions of the given ODD (e.g., sudden storm, damaged road infrastructure, etc.) creates a situation that may result in an immediate risk of collision.

Failure situations involve those in which the ADS or another vehicle system experiences a fault or failure that removes or reduces the ADS's ability to perform the DDT, such as sensor, computer, or propulsion system failure.

1.3 Application of Rules of the Road

One of the open questions in the scenario-based testing of ADS remains the definition of the pass or acceptance criteria. The UNECE Framework document on automated/autonomous vehicles [8] mentions that the ADS need to *"ensure compliance with road traffic regulations"*. It is challenging to test against this requirement in the absence of "codified rules of the road".

Furthermore, an approach is proposed to create a natural language description and machine-readable description of the codified rules of the road, which can be used by:

- Natural language description: regulators or type approval authorities
- Machine readable: ADS developers, OEMs, suppliers, etc., for simulation-based testing purposes and to identify gaps and contradictions in the rules

If one compares the scope of ODD and the content of current *"rules of the road for human drivers"* (e.g., the UK's Highway Code [6]), a large overlap of scenery aspects and environmental condition aspects can be observed. It is therefore plausible to follow the ODD-based approach and the ODD taxonomy to model the environmental and scenery aspects of the *"rules of the road"*, while competencies can be divided into ego (vehicle under test) behaviours and actor behaviours, as shown in Fig. 2. Any rule of the road can be classified into two categories:

- Doing some behaviour somewhere
- NOT doing some behaviour somewhere

While doing or not doing can be defined as part of ADS's behavioural competencies, *"somewhere"* could be considered as an "operating condition" or part of the ODD definition.

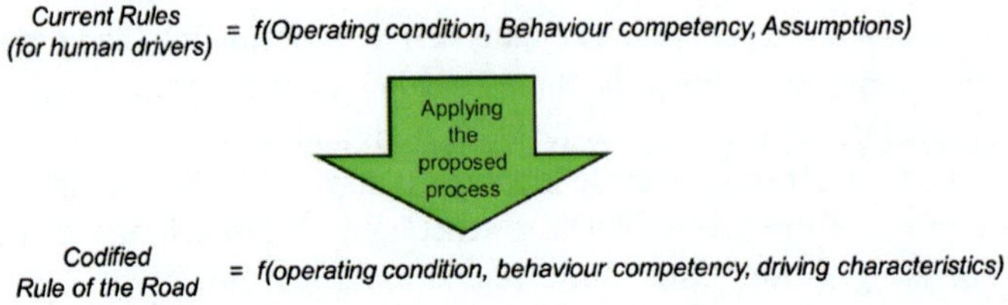

Fig. 2. Approach for rules of the road codification

Every test scenario definition will have ODD and behaviour competency attributes defined. The same can be found in road definition rules. Therefore, it is possible to map every scenario to a corresponding rule(s) of the road using ODD and behaviour tags or labels within a scenario catalogue, as shown in Fig. 3. This approach would allow the test engineer to map each scenario to a corresponding rule (or set of rules), and these can then serve as the pass criteria during the scenario-based testing approach. This approach can enable engineers to show traffic rule compliance by making the rules of the road verifiable.

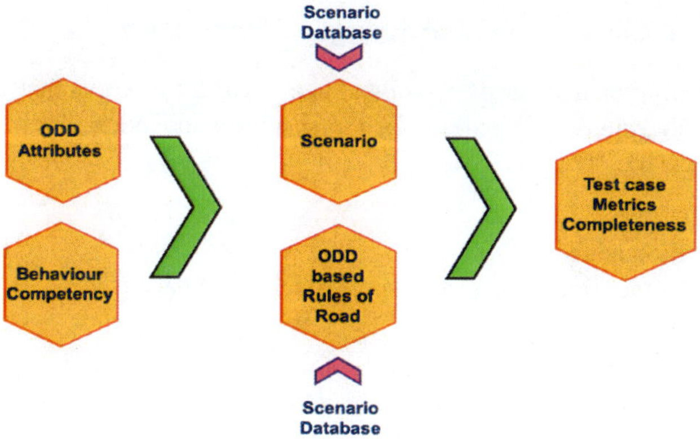

Fig. 3. Codified rules of the road for scenario-based testing

2 ODD-Based Safety Assurance Framework (ODD-SAF)

The elements introduced in the previous paragraphs are all part of the overall ODD Safety Assurance Framework. This can be summarised by considering the interaction of the following key elements, as shown in Fig. 4:

- Behavioural Competencies and Scenarios Identification
- Competencies and Scenarios Mapping: Functional to Concrete
- Assumptions
- Performance Evaluation

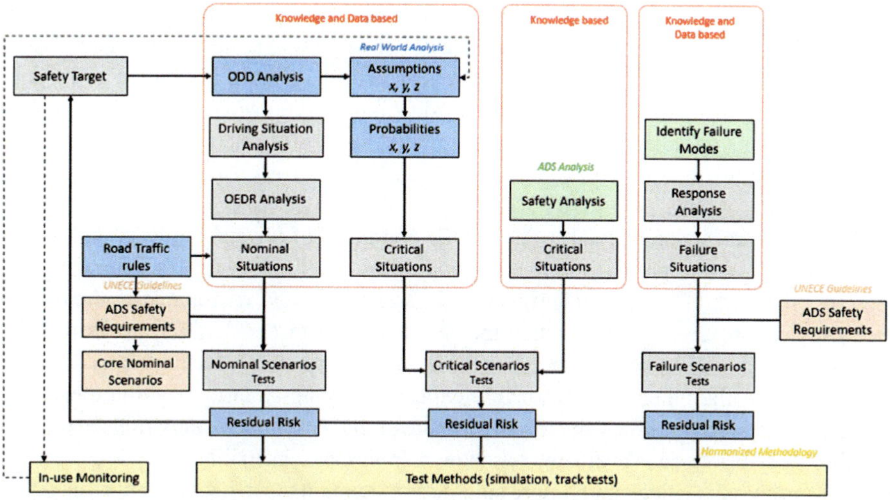

Fig. 4. ODD-SAF process overview

2.1 Behavioural Competencies and Scenarios Identification

The ODD-SAF approach suggests a series of analytical frameworks that could help to derive measurable criteria appropriate for the specific application. These frameworks are divided into:

- ODD analysis
- Driving Situation analysis
- Objects and Events Detection and Response (OEDR) analysis

2.1.1 ODD Analysis

This analysis represents the first step with the aim to identify ODD characteristics. An ODD may consist of stationary physical elements (e.g., physical infrastructure), environmental conditions, dynamic elements (e.g., reasonably expected traffic levels and composition, vulnerable road users) and operational constraints to the specific ADS application. Various sources provide useful guidance for precisely determining the elements of a particular ODD and their format definition [1–4]. As part of this activity, the level of detail of the ODD definition using ODD attributes also need to be established.

2.1.2 Driving Situation Analysis

In driving situation analysis, the behaviours of other road users are reasonably expected, and presence of ODD roadway characteristics are explored in more detail by mapping actors with appropriate properties and defining interactions between the objects.

The NHTSA [9] provides an example of this analysis in Table 1, where static and dynamic behaviours of other objects (including other road users) the ADS is reasonably expected to encounter within the ODD are described. In the case of vehicles, this includes dynamic behaviours such as acceleration, deceleration, cut-ins; for pedestrians, crossing the road or walking on sidewalk. Some of these behaviours may involve nominal situations (e.g., lead vehicle deceleration at a rate reasonably expected in light of traffic and other circumstances within the bounds of physical limitations) while others may involve critical situations (e.g., sudden cut-ins or unpredictable pedestrian or cyclist behaviour, or other behaviours that may violate local traffic laws, such as crossing a road outside a designated crosswalk).

The behaviour of other road users and the condition of physical objects within the ODD may fall at any point along the continuum of likelihood. For example, deceleration by other vehicles may range from what is expected and reasonable in the traffic circumstances, to unreasonable but somewhat likely rapid deceleration, to extremely unlikely (e.g., a sudden cut-in combined with full braking on a clear high-speed road). The analysis of the ODD and reasonably expected driving situations within the ODD should make distinctions that include an estimate of the likelihood of situations to ensure the ADS's performance is evaluated based on response to reasonably likely occurrences involving nominal, critical, and failure situations but not on the expectation that the ADS will avoid or mitigate the most extremely unlikely occurrences.

Table 1. Example of static and dynamic properties of objects and other road users

Objects	Events / Interactions
Vehicles	Lead vehicle decelerating, lead vehicle stopped, lead vehicle accelerating, changing lanes (frontal/side), cutting in (adjacent), turning (frontal), encroaching adjacent vehicle (frontal/side), entering roadway (frontal/side), cutting out (frontal)
Pedestrians	Crossing road – inside crosswalk, crossing road – outside crosswalk, walking on sidewalk/shoulder
Bicycle	Riding in lane, riding in adjacent lane, riding in dedicated lane, riding on sidewalk/shoulder, crossing road – inside crosswalk, crossing road – outside crosswalk
Animals	Static in lane, moving into/out of lane, static/moving in adjacent lane, static/moving on shoulder
Debris	Static in lane
Other dynamic objects (e.g. shopping carts)	Static in lane, moving into/out of lane
Traffic Signs	Stop, yield, speed limit, crosswalk, railroad, crossing, school zone
Vehicle signals	Turn signals

2.1.3 OEDR Analysis: Behavioural Competencies Identification

Once the objects and their reasonably expected behaviours have been identified, it is possible to map the appropriate ADS response, which can be expressed as a behavioural competency. The detailed response is derived from more general and applicable functional requirements defined by regulators. The acceptable ADS response will vary depending on whether the driving situation involves nominal, critical, or failure characteristics. The outcome of the analysis is a set of behaviour competencies that can be applied to the events characterising the ODD. NHTSA [9] provides a qualitative example of a matching event – response in Table 2. The combination of objects, events, and their potential interaction - as a function of the ODD - constitute the set of nominal or critical situations pertinent to the ADS under analysis.

Table 2. Example of elementary behaviour competencies for given events

Event	Response
Lead vehicle decelerating	Follow vehicle, decelerate, stop
Lead vehicle stopped	Decelerate, stop
Lead vehicle accelerating	Accelerate, follow vehicle
Lead vehicle turning	Decelerate, stop
Vehicle changing lanes	Yield, decelerate, follow vehicle
Vehicle cutting in	Yield, decelerate, stop, follow vehicle
Vehicle entering roadway	Follow vehicle, decelerate, stop
Opposing vehicle encroaching	Decelerate, stop, shift within lane, shift outside of lane
Adjacent vehicle encroaching	Yield, decelerate, stop
Lead vehicle cutting out	Accelerate, decelerate, stop
Pedestrian crossing road – inside crosswalk	Yield, decelerate, stop
Pedestrian crossing road – outside crosswalk	Yield, decelerate, stop
Cyclist riding in lane	Yield, follow
Cyclist riding in dedicated lane	Shift within lane
Cyclist crossing road – inside crosswalk	Yield, decelerate, stop
Cyclist crossing road – outside crosswalk	Yield, decelerate, stop

2.1.4 Scenarios Identification

Two types of methods may be used for scenario identification:

- Knowledge-based
- Data-based

A knowledge-driven scenario identification approach utilises domain specific (or expert) knowledge to identify hazardous events systematically and create scenarios.

Analytical hazard-based methods (e.g. The Systems Theoretic Process Analysis, STPA) can be used to analyse the characteristics of the ADS architecture and identify system failures and hazardous situations. Other knowledge-based methods include the formal analysis approach with highway code rules to identify scenarios; or the formal representation of the ODD and ADS behaviour competencies for scenario generation. Furthermore, existing scenarios defined in standards, regulations, or guidelines can also be utilised for the ADS testing.

A data-driven approach utilises the available data to identify and classify occurring scenarios. Accident datasets can be analysed to identify accident hotspots and scenario parameters, which contribute to causation of accidents carrying high levels of severity. Additionally, anonymised real-world data can also be analysed to identify the trends in near-miss events.

Figure 5 highlights the interaction between knowledge and data-based methods for scenario identification.

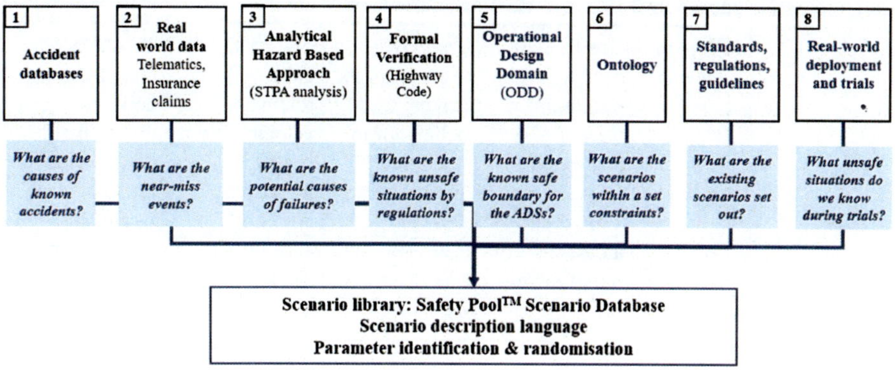

Fig. 5. Various data-based (DB) and knowledge-based (KB) scenario generation methods

2.2 Competencies and Scenarios Mapping: Functional to Concrete

Once the behavioural competencies and scenarios are identified, it is of utmost importance to link them appropriately. This can be done by considering the three broad categories of driving situations that may be encountered in performance of the DDT, such as nominal, critical, and failure.

2.2.1 Nominal Situations

In these situations, ADS competencies can often be specified by applying traffic laws of the country where the ADS is intended to operate, as well as by applying general safe driving principles for situations not addressed adequately by current traffic laws for human drivers. Examples of such competencies may include adherence to legal requirements to maintain a safe distance from vehicles ahead; provide pedestrians the right of way; obey traffic signs and signals; and more.

Some nominal competencies (e.g., safe merging, safely proceeding around road hazards) may not be explicitly articulated or mandated by traffic laws. In some instances, traffic laws may provide wide discretion for the driver to determine the safest response to a particular situation, such as responding to adverse weather conditions; therefore without sufficient specificity to provide a clear basis for defining a competency. As such, the application of models involving safe driving behaviour may be needed in addition to reference to codified rules of the road in developing more granular behavioural competencies for nominal driving situations.

Safe driving behaviour can be assessed via a suite of nominal scenarios pertinent to the ADS under analysis, as shown in Table 3.

Table 3. Example of scenario and competencies mapping in nominal situation

ODD Element	Driving behaviour	Traffic Rule	ADS Safety Requirement	Behaviour competency	Test Scenario
Bicycle	Riding in Lane	Drivers will need to use a minimum passing distance for bicycles of 1.5m in urban areas, and 2m out of town	The ADS shall adapt its driving behaviour in line with safety risks	The ADS ensures relative velocity during passing manoeuvre does not exceed [30] km/h	The ADS travels between [30–50]km/h on the centre line of its lane A cyclist travels in the same direction as the ADS between [10–20] km/h, [0.2–1] m away from the lane edge
			The ADS shall comply with traffic rules	The ADS shifts in lane to pass by cyclist with 1.5.m lateral distance	
			The ADS shall adapt its driving behaviour to the surrounding traffic conditions (e.g. by avoiding disruption to the flow of traffic)	The ADS crosses the centre lane marking to ensure the safe passing distance is not violated	
			The ADS shall interact safely with other road users	The ADS activates the turn signal if the centre lane marking is crossed	

2.2.2 Critical Situations

The development of these competencies requires analysis of what constitutes such unreasonable behaviour by other road users (ORUs) and/or a sudden change of the operating conditions that are not reasonably foreseeable and what constitutes an appropriate ADS response to avoid or mitigate the imminent crash. Additionally, it is also important to identify the occurrence of unplanned emergent behaviour in critical situations.

Analysis of the first type may be based on a variety of methodologies, including IEEE 2846–2022 [10] (which offers guidance on what behaviours by other road users are reasonably foreseeable) and other models of reasonable driving behaviour. Analysis of the second factor may be based on various models of acceptable human driving behaviour in crash imminent situations.

Besides, hazard identification methods such as STPA, which analyse the system design for functional and operational insufficiencies can help identify the occurrence of emergent behaviour, which may lead to critical situations [11, 12]. STPA is based on system engineering and considers system safety as a control problem. Therefore, breaches of control laws (or constraints) cause accidents. The analysis can be summarised by the following four steps:

- Identify system-level hazards
- Creation of system control structure
- Identify unsafe control actions (UCAs)
- Identify causal factors

Furthermore, the UCAs and causal factors can be parameterized to derive test scenarios and pass/fail criteria.

In the example depicted in Table 4, the identified hazard *"ADS does not maintain safe distance from lead motor vehicle"* is linked to the relevant unsafe control action *"braking demand is not provided"* and to the potential causal factors *"undetected/misclassified object"* or *"incorrect sensor fusion results"*. The UCA and the causal factors can then be parameterized to generate a critical scenario.

Development of behavioural competencies for critical driving situations faces several challenges. No consensus exists on the appropriate models for the behaviour of ORUs or appropriate responses by the ADS to unreasonable ORU behaviours that make a crash imminent.

2.2.3 Failure Situations

Failure situations involve those in which the ADS or another vehicle system experiences a fault or failure that removes or reduces the ADS's ability to perform the DDT, such as sensor, computer, or propulsion system failure.

In developing the behavioural competencies appropriate for failure situations, the objective is to describe the ability of the ADS to detect and respond safely to specific types of faults and failures. Depending upon the nature and extent of the fault or failure, the responses can include identifying a minor fault for immediate repair after trip completion; responding to a significant fault with restrictions (such as limp-home mode) for the remainder of the trip; or responding to major failures by achieving a minimal risk condition. Communication of the fault or failure condition to vehicle users may also be a desirable ADS behavioural competency.

Different methods are available in literature and include the application of Failure Modes and Effects Analysis (FMEA) [13] for which an example is given in Table 5.

An FMEA can generally include the following steps:

- Identify potential failure modes
- Identify potential causes and effects of those failure modes
- Prioritise the failure modes based upon risk

Table 4. Example scenario and competencies mapping in critical situation

Losses	Hazards	Unsafe Control Action	Loss scenario	Causal factors	Test behaviour	Test Scenario
Collision with object outside the vehicle	ADS does not maintain a safe distance from the lead motor vehicle	Braking demand is not provided	Object in vehicle trajectory is not detected	Undetected/ misclassified object; Obscured object; Incorrect sensor fusion result	The ADS is following behind a lead vehicle, with the headway set by the ADS. The lead vehicle decelerates at the max assumed rate depending on the weather conditions	Lead vehicle decelerated to turn [right/left] or travel straight on a [mini / large] roundabout
			Object is not considered to be in the vehicle trajectory	Localisation issues leading to incorrect positioning of ego vehicle or object		Lead vehicle decelerated whilst shifting lane to avoid a [static object/other road user]

– Identify appropriate corrective actions or mitigation strategies

Table 5. Example of FMEA

Behaviour failure	Effects
Fail to maintain lane	Impact adjacent vehicle or infrastructure
Fail to maintain safe following distance	Impact lead vehicle
Fail to detect and respond to manoeuvres by other vehicles	Impact lead or adjacent vehicles
Fail to detect relevant obstacles in or near lane	Impact obstacles
Fail to identify ODD/OEDR boundary	Operate outside of ODD/OEDR capabilities

For each of the behaviour failures and consequential effects listed, the manufacturer put in place relevant strategies when developing the ADS. By way of example, these might include a Fail-Safe strategy, where the primary goal is to rapidly achieve a Minimal Risk Condition (MRC) via transition to fallback-ready user or safely stop in lane of travel; or a Fail-Operation strategy, allowing the ADS to continue to function in a degraded mode for a limited duration.

When applying the failure scenarios, the objective is to assess the ability of the ADS to comply with requirements highlighted by the UNECE Guidelines [8], as outlined in Table 6.

Table 6. Example of scenario and competencies mapping in failure situation

Failure Type	Failure Mode	Potential Cause	Response	ADS Safety Requirement	Test scenario	Pass / Fail criteria
Perception	Fail to identify ODD boundary	Failure to detect ODD attribute e.g. heavy rain/fog	Safely stop in lane of travel	The ADS shall be able to detect the ODD and predict when the ADS is about to leave the ODD	The ADS operates beyond the predicted ODD	The ADS detects the ODD conditions are not met and issues a minimal risk manoeuvre
				When the system detects that it is difficult to continue in the ADS mode, it shall be able to transfer to a minimal risk condition through a minimal risk manoeuvre		The minimum risk manoeuvre should not cause the vehicle to decelerate greater than [4]m/s^2

2.3 Assumptions: Logical to Concrete Behavioural Competencies

Concrete behavioural competencies depend on the specific situations the ADS encounters - on a reference behaviour that is deemed appropriate for a human driver or a technical system, and on assumptions about vehicle and other road users' behaviours.

Assumptions concerning the actions of other road users may need to account for cultural differences in driving styles in different geolocations, making it impracticable to harmonise these assumptions across different domains. Therefore, evidence should be provided to support the assumptions made. Existing standards such as IEEE 2846–2022 [10] provide a set of assumptions to be considered by ADS safety-related models for an initial set of driving situations.

Additionally, several other tools including data collection campaigns performed during the development phase; real-world accident analysis and realistic driving behaviour evaluations; constraint randomisation; and, Bayesian optimisation among others can be used to inform values for such assumptions.

2.4 Performance Evaluation

The UNECE Framework on Automated Vehicles [8] requires that *"when in automated mode, the automated/autonomous vehicle should be free of unreasonable safety risks to the driver and other road users and ensure compliance with road traffic regulations."*

Translating this concept into performance criteria is not immediate, however, one can refer to the different classification of driving situations, as defined above.

Given the nature of nominal situations - those in which behaviour of other road users and the operating conditions of the given ODD are reasonably foreseeable and no failures occur that are relevant to the ADS's performance of the DDT - it is expected that the ADS would be capable of handling them without any resulting collision.

According to the UNECE requirements for "*failsafe response*", in failure situations the ADS is expected to recognise faults/failures within the system and manage safety-critical situations in a safe manner, either implementing a fail-safe strategy or operating in a degraded mode.

On the other hand, defining performance criteria in critical situations, such as where others are at fault and behaving unforeseeable and the collision might potentially not be prevented, need to be further analysed. No consensus exists on the appropriate responses by the ADS to unreasonable ORU behaviours that make a crash imminent. In this case, it is recognised that the ADS may not be able to avoid a collision, so the ADS performance needs to be compared with safety model performance to set the threshold between where avoidance is required and where it is not feasible, and if mitigation may be possible [14].

3 Conclusions

The safety assurance of an ADS poses challenges that cannot be addressed exclusively with the application of prescriptive requirements and with their assessment through a series of repeatable tests, which is common in the traditional certification regimes.

The holistic ODD Safety Assurance Framework (ODD-SAF), as proposed, plays a key role in determining operational boundaries and defining the ADS behavioural competencies required in nominal, critical and failure situations. On these bases, the framework allows the identification of scenarios where these competencies can be assessed by both developers and certification authorities.

Whereas collision avoidance is the key aspect in nominal situations, it is recognised that the ADS may not be able to avoid a collision in cases where others are at fault and/or a sudden and not reasonably foreseeable change of the operating conditions creates a situation that may result in an immediate risk of crash.

Despite the fact that, at present, there is no consensus on the expectations of the ADS responses in these critical situations, safety models are proposed in literature and based on a reference behaviour that is deemed appropriate for a human driver or a technical system, and on assumptions about vehicle and other road users' behaviours.

However, further work still needs to be done in the space of critical situations to allow for a better understanding of what is a level of safety that is deemed appropriate and aligned with the expectations of regulators worldwide.

Annex I – Use-Case for Nominal, Critical and Failure Situations Mapping

(See Tables 7, 8 and 9)

Table 7. Example of Scenario and Competences Mapping, Nominal Situation

ODD Element	Driving Behaviour	Traffic Rule	ADS Safety Requirement	Behavioural competency	Assumption	Test scenario	Pass/Fail criteria
Bicycle	Riding in Lane	Drivers will need to use a minimum passing distance for bicycles of 1.5m in urban areas, and 2m out of town	The ADS shall adapt its driving behaviour in line with safety risks	The ADS ensures relative velocity during passing manoeuvre does not exceed [30] km/h	Bicycle $V_{long} = 4.3$ [SD 0.57] m/s	The ADS travels between [30–50]km/h on the centre line of its lane. A cyclist travels in the same direction as the ADS between [10–20]km/h, [0.2–1]m away from the lane edge	The relative speed between the ADS and cyclist shall not exceed [30]km/h
			The ADS shall comply with traffic rules	The ADS shifts in lane to pass by cyclist with 1.5.m lateral distance			The passing distance between the cyclist and the ADS is not less than 1.5m
			The ADS shall adapt its driving behaviour to the surrounding traffic conditions (e.g. by avoiding disruption to the flow of traffic)	The ADS crosses the centre lane marking to ensure the safe passing distance is not violated			The ADS may cross the centre lane marking to ensure the passing distance is not violated
			The ADS shall interact safely with other road users	The ADS activates the turn signal if the centre lane marking is crossed			The ADS activates the turn signal if the centre lane marking is crossed

Table 8. Example of Scenario and Competences Mapping - Critical Situation

Losses	Hazards	ODD	Control structure	Control action	Unsafe Control Action	Loss scenario	Causal factors	Assumption	Test behaviour	***Test Scenario***	***Pass / Fail criteria***
Collision with object outside the vehicle	The ADS does not maintain a safe distance from the lead motor vehicle	Urban; Day and Night; All weather conditions	Level 4 (no driver); Sensors: LIDAR, RADAR, Camera; Actuation: Brake, Accelerator, Steering; No V2X	Request braking command	Braking demand is not provided	Object in vehicle trajectory is not detected	Undetected/ misclassified object; Obscured object; Incorrect sensor fusion result	Lead vehicle deceleration = 7.0 [SD 2.3] m/s (dry) = 4.4 [SD 1.0] m/s (wet)	The ADS is following behind a lead vehicle, with the headway set by the ADS. The lead vehicle decelerates at the max assumed rate depending on the weather conditions	Lead vehicle decelerated to turn [right/left] or travel straight on a [mini / large] roundabout	The ADS avoids a collision with the lead vehicle
						Object is not considered to be in the vehicle trajectory	Localisation issues leading to incorrect positioning of ego vehicle or object			Lead vehicle decelerated whilst shifting lane to avoid a [static object/other road user]	

Table 9. Example of Scenario and Competences Mapping - Failure Situation

Failure Type	Failure Mode	Potential Cause	Response	ADS Safety Requirement	Assumption	Test scenario	Pass / Fail criteria
Perception	Fail to identify ODD boundary	Failure to detect ODD attribute e.g. heavy rain/fog	Safely stop in lane of travel	The ADS shall be able to detect the ODD and predict when the ADS is about to leave the ODD	N/A	The ADS operates beyond the predicted ODD	The ADS detects the ODD conditions are not met and issues a minimal risk manoeuvre
				When the system detects that it is difficult to continue in the ADS mode, it shall be able to transfer to a minimal risk condition through a minimal risk manoeuvre			The minimum risk manoeuvre should not cause the vehicle to decelerate greater than [4]m/s^2

References

1. BSI: Operational Design Domain (ODD) taxonomy for an automated driving system (ADS) – Specification. The British Standards Institution, BSI PAS 1883 (2020)
2. SAE ITC: AVSC Best Practice for Describing an Operational Design Domain: Conceptual Framework and Lexicon (2020)
3. ASAM: OpenODD Project Proposal (2020)
4. ISO: ISO/DIS 34503 - Road Vehicles - Test scenarios for automated driving systems - Taxonomy for operational design domain (2022)
5. ASAM OpenLABEL1.0. https://www.asam.net/standards/detail/openlabel/
6. The Highway Code. https://www.gov.uk/guidance/the-highway-code
7. SAE ITC: AVSC Best Practice for Evaluation of Behavioural Competencies for Automated Driving System Dedicated Vehicles (ADS-DVs) (2021)
8. UNECE: Revised Framework document on automated/autonomous vehicles (2019) https://unece.org/DAM/trans/doc/2019/wp29/ECE-TRANS-WP29-2019-34-rev.1e.pdf
9. NHTSA: A Framework for Automated Driving System Testable Cases and Scenarios (2018)
10. IEEE Standard for Assumptions in Safety-Related Models for Automated Driving Systems. in IEEE Std 2846–2022, vol., no., pp.1–59, 22 April 2022, https://doi.org/10.1109/IEEESTD.2022.9761121
11. ISO: ISO/PAS 21448 - Road Vehicles - Safety of the Intended Functionality (2022)

12. UL: ANSI/UL 4600 - Standard for Safety for the Evaluation of Autonomous Products (2023)
13. ISO, ISO 26262 - Road Vehicles - Functional Safety (2018)
14. Mattas K., et al.: Driver models for the definition of safety requirements of automated vehicles in international regulations. Application to motorway driving conditions, Accident Analysis & Prevention, vol. 174 (2022)

Automated Vehicle Testing & Data Collection Efforts

Xin Xia[1], Stephane Dreher[2], Jiaqi Ma[1], Stefan de Vries[3], Guoyuan Wu[4], and Chris Schwarz[5](✉)

[1] University of California, Los Angeles, CA, USA
{x35xia,jiaqima}@g.ucla.edu
[2] ERTICO, Avenue Louise 326, 1050 Brussels, Belgium
s.dreher@mail.ertico.com
[3] IDIADA, Tarragona, Spain
Stefan.DeVries@idiada.com
[4] University of California, Riverside, CA, USA
guoyuan.wu@ucr.edu
[5] The University of Iowa, Iowa City, IA, USA
chris-schwarz@uiowa.edu

Abstract. We summarize presentations from an international slate of speakers at the 2022 Automated Road Transportation Symposium on the topic of testing and data collection efforts for automated vehicles (AV), referred to throughout the chapter as automated driving systems (ADS), connected automated vehicles (CAV), cooperative driving automation (CDA), and connected, cooperative and automated mobility (CCAM). Projects covered in this chapter include an ADS Data Acquisition & Analytics Platform developed at UCLA, the European data sharing initiatives linked to the EU funded projects ARCADE and FAME, the Data For Road Safety (DFRS) initiative and the two EU funded projects HEADSTART and SUNRISE focusing on AV safety assessment.

Keywords: Automated vehicle data · Data sharing framework · Data standardization

1 Introduction

There are several efforts underway or about to begin to develop shared frameworks for testing automated vehicles and to streamline the process of collecting and processing data. These efforts are designed to meet needs throughout the process of testing and certifying automated driving systems (ADS) so that their operational performance and safety can be assured. Work is focused both on the technical problem of collecting data as well as the format and content of the resulting data. A few of these were highlighted during a breakout session at the 2022 Automated Road Transportation Symposium (ARTS).

The session was organized to highlight efforts on the scale of national and international bodies as opposed to proprietary tools developed by individual companies. The

G. Meyer and S. Beiker (Eds.): ARTSymposium 2022, LNMOB, pp. 152–162, 2023.
https://doi.org/10.1007/978-3-031-34757-3_12

goals of the session were to present new project updates that offered international perspectives, presented cutting edge work, projected out to the coming year, and informed the audience about shared frameworks, datasets, and tools.

2 Summary of the Presentations

Invited speakers shared five presentations that cut across international borders and applications. Three of the presentations are summarized in this chapter.

Professor Jiaqi Ma of UCLA presented work on the ADS Data Acquisition & Analytics Platform that he participated in with the Transportation Research Center (TRC). This platform is a US initiative that intends to streamline the data flow, from collection to processing, for multiple ADS-equipped vehicles.

Dr. Stephane Dreher of ERTICO presented two initiatives from the European Union. ARCADE is a Coordination and Support Action (CSA) aimed at building consensus across many stakeholders and is funded through July 2022. A test data sharing framework was developed that standardizes data and metadata descriptions while trying to protect data security and privacy. The Data for Road Safety ecosystem is a cross-border, cross-brand, public private cooperation managed by ERTICO-ITS Europe. Information is collected on events such as accident areas, debris on road, reduced visibility, work zones, and others.

Stefan de Vries of IDIADA presented two European initiatives called HEADSTART and SUNRISE. HEADSTART defines testing and validation procedures for Connected, Cooperative and Automated Mobility (CCAM) functions, including enabling technologies, cross-linking simulations with field tests, and validating safety and security performance. The Safety Assurance Framework for Connected, Automated Mobility Systems (SUNRISE) aims to develop a harmonized and scalable safety assurance framework that fulfills the needs of different automotive stakeholders and is scheduled to run from September 2022 through August 2025.

2.1 ADS Data Acquisition and Analytics Platform

With new disruptive connected automated vehicles (CAVs) technology looming, researchers and engineers need to understand the benefits coming from this technology and, in the meantime, are preparing for the future challenges that these vehicles will encounter, and potentially cause to traffic.

Automated driving systems (ADS) in vehicles equipped with multi-modal sensors (e.g., LiDAR, camera, RTK-corrected global navigation satellite system (GNSS)) for vision provide advanced sensor data that can be used to perceive the traffic environments. An enormous amount of information can be extracted to understand new mixed traffic performance, such as ADS operational safety, the interaction between CAVs, and other traffic, for instance, traffic oscillations, heterogeneity, car-following behavior and lane change behavior [1, 2]. Therefore, there is a need for an ADS data acquisition and processing platform to process the advanced sensor data in ADS-equipped vehicles or CAVs to obtain surrounding vehicles' trajectory data for ADS and transportation communities.

The Next Generation Simulation program (NGSIM) dataset [3] which is collected by processing the video from cameras installed on infrastructure is commonly used. Traditional datasets are used mainly for analyzing human-driven vehicle behavior and traffic flow. To better understand the influence on traffic safety, reducing traffic congestion, and reducing energy consumption by involving individual ADS or CAV in traffic, efforts from both industry and academia have been made and datasets such as KITTI [4], and Waymo Open Dataset [5] have been proposed. To further investigate the impact of CAVs on traffic, a team from UCLA has published OPV2V [6] consisting of raw sensor data from LiDAR, cameras, and GNSS/IMU in multiple CAVs. These human-labeled datasets for both AVs and CAVs can be leveraged to extract the objects' (surrounding vehicles) trajectory.

However, it is not realistic to always use labor to label the datasets and to analyze the influence whenever the software updates since manual labeling is very expensive. Therefore, a platform is needed that can collect sensor data from advanced sensors and process them to obtain the objects' trajectories that the transportation community needs.

CAV sensors may include 3D-LiDARs, cameras, radars, and GNSS. 3D-LiDAR has the capability to detect objects within a range at high accuracy. When there are multiple CAVs, the LiDAR data on different vehicles can be fused allowing more surrounding vehicles to be detected. In this way the sensor detection range of each vehicle is virtually extended, which is beneficial to construct the traffic flow on a larger scale. However, to the best of our knowledge, there is neither research that utilizes the 3D-LiDAR sensors on the equipped vehicle nor works which leverage the advanced sensors on multiple-CAVs to extract the surrounding vehicles' trajectories. This work is trying to fill these gaps and propose a holistic and systematic platform that is able to collect multi-modal sensors from multiple CAVs and to process them to extract, reconstruct, and evaluate the trajectories.

Figure 1 shows the ADS data acquisition and analytics platform (ADAAP) proposed by the UCLA Mobility Lab. Both the sensor data from CAVs and infrastructures can be collected and processed. CAVs are commonly equipped with multiple advanced sensors such as GNSS/IMU integration system, LiDARs, cameras, radars, and onboard sensors including wheel speed sensors.

When data from multiple CAVs are combined, the benefits of cooperative perception can be realized. Figure 2 shows a section of freeway viewed in Google earth (Fig. 2a), as a vector map (Fig. 2b), and with sensor data (Fig. 2c). The operating speed is about 112km/h. The red point cloud is from the LiDAR in CAV1 and the white point cloud is from CAV2. While SV2 is too far behind CAV2 to be detected, it is picked up by the LiDAR of CAV1.

Thanks to this cooperative perception, the sensing area of each CAV has been expanded and more SVs can be detected which means the output of the data processing framework provides the potential to analyze complex interactions between CAV and multiple SVs.

2.2 Data Sharing Initiatives in the European Union

Automated Vehicles Data sharing challenges in European Union (EU) are mostly related to 1) limited exchange of data between different testing activities, 2) an increased need

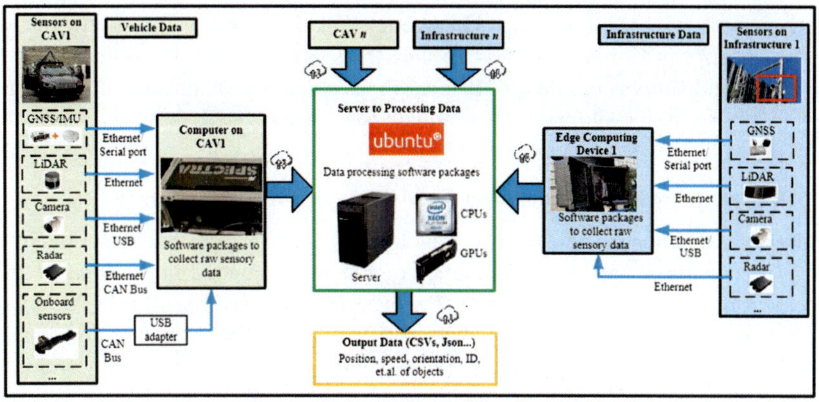

Fig. 1. ADS data acquisition and analytics platform. Sensor data from CAVs and infrastructures can be acquired by the computer on CAV or edge computing device at the infrastructure.

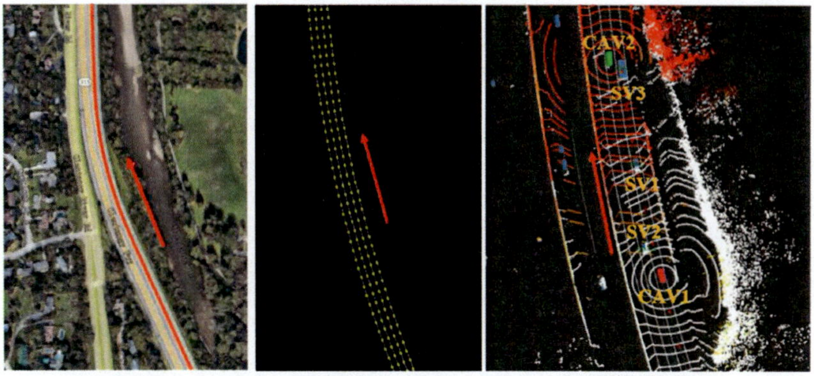

Fig. 2. Visualization of test scenario, vector map, and object detection and tracking. The red arrow in these three sub-figures shows the moving direction of the vehicle. In (c), the colored and white point clouds are from the LiDARs.

to access non-competitive data for safety validation, impact assessment or training of automated driving functions and vehicles, and 3) public-private collaboration to address legislation requiring the provision of safety related data for public good.

Many diverse ADS testing and demonstration activities are being carried out by EU funded projects but also at national, regional or cities level across Europe. The majority of funding is often used to collect data, leaving only a small part for the analysis. The absence of proper or harmonised data description and the lack of funding after the end of projects constitute a barrier for the maintenance, exchange and reuse of test data.

The European Data Strategy [7] highlights that data sharing has not taken off at sufficient scale due to lack of economic incentives including fear of losing competitive edge, lack of trust in agreements being followed, fear of misappropriation of the data by third parties, and a lack of legal clarity on who can do what with the data. A series of legislative efforts have been developed in recent years in the frame of the European Data Act by the European Commission to facilitate the sharing of data between private and public sector and support the flow of data across sectors and countries. This legislative context is driving the development of adequate frameworks at EU level.

2.2.1 Data Sharing Framework and FAME Project Test Data Space

To address the challenges of sharing Vehicle Test Data, a Data Sharing Framework (DSF) [8] has been set up through a series of European Funded projects since 2008, complementing the FESTA methodology for Field Operational Tests (FOTs) which was developed in the European FESTA project [9]. The Framework was first developed in the EU funded FOT-Net [10] projects and extended to automated driving within the CARTRE [11] and ARCADE [12] support actions coordinated by ERTICO. The aim of the framework is to support projects and organizations by giving hands-on recommendations in different topics, all important to enable data re-use or sharing. The framework is focused on FOTs or Naturalistic Driving Studies (NDS), but has proven to be useful, with some adaptions, to other domains.

The CARTRE project released an update of the FOT-Net framework in 2019 following the implementation of the European General Data Protection Regulation [13]. In February 2021, ARCADE organised a Workshop in Brussels, Belgium, to discuss bottlenecks and directions for sharing of data related to Connected, Cooperative and Automated Mobility (CCAM) [14]. An increased focus is required on data formats to make them interoperable (e.g. Common Data format developed by the L3Pilot project [15], scenario database, edge cases, simulation specific formats,...) and on data sharing services (Gaia-X [16], International Data Space Association [17], and the future Mobility Data Space). The ongoing project PrepDSpace4Mobility funded by the European Commission under the Digital Europe programme is preparing an inventory of Data ecosystems in Europe [18] which shall ultimately provide a better understanding of the current landscape of such services.

The EU funded project FAME [19], which follows up on ARCADE and started in July 2022, will develop a European Framework for Testing of CCAM on Public Roads, where data sharing is one of the topics covered. The current collection of knowledge and methodologies (top of Fig. 3) available in the EU-wide Knowledge Base [20] set up as part of the ARCADE project, includes existing components (FESTA and Data Sharing Framework). FAME will further consolidate best practises, develop checklists, recommendations and templates for describing testing activities for policy and legal processes, ethics and data sharing. The development has started with the taxonomy to ensure common terminology.

2.2.2 Data for Road Safety (DFRS)

A good example of a successful public private cooperation to facilitate data sharing in Europe is the Data for Road Safety (DFRS) [21] initiative that was set up to improve road safety across the European Union to support the achievement of Vision Zero [22] and reducing the number of road fatalities and accidents. It does not focus directly on Automated Driving but the data ecosystem that it built supports it.

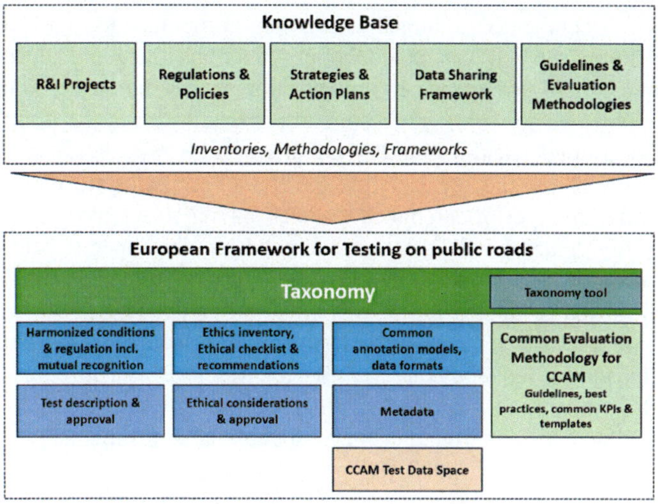

Fig. 3. Current content of the Knowledge Base (top) and main components of the European Framework for Testing on Public Roads developed in the FAME project (bottom)

The DFRS initiative was set up in 2017 under the form of a Data Task Force in the frame of the High Level Meeting on Connected and Automated Driving in Amsterdam [23]. The participating Member States and the industry set up this partnership to define initial steps for deployment of data sharing for traffic safety related data in real life situations. The group decided its scope to be Safety Related Traffic Information (SRTI) because of the high societal value, technological readiness and commitment from the stakeholders.

DRFS takes a transparent and cooperative approach. A key element in the setup of this ecosystem has been the definition of levels of data and roles for the participating stakeholders. Three types of data are considered in the ecosystem (Fig. 4).

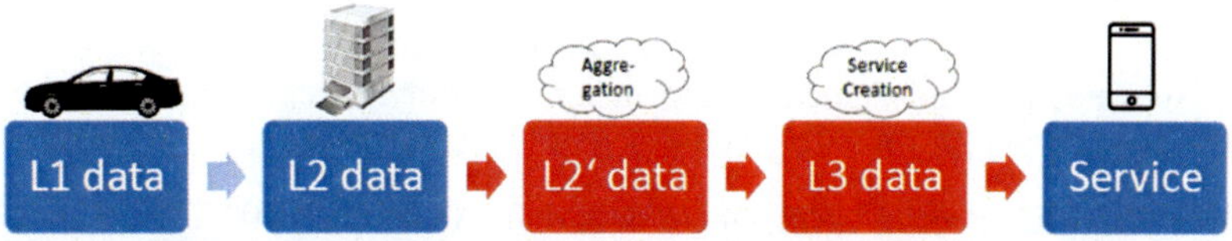

Fig. 4. Levels of Data of the DFRS SRTI ecosystem. L2 is raw data from vehicles or other sources. L2' data is enriched by cross-referencing/cleansing data. L3 data is aggregated and composed of "Road Safety Related Minimum Universal Traffic Information" or "SRTI"

2.3 Harmonized CCAM Methodologies: HEADSTART and SUNRISE Projects

Data collection and generation of scenarios is essential for the validation of Automated Driving functions. The EU funded HEADSTART project [24] coordinated by IDIADA, which ended in December 2021, aimed at defining harmonized testing and validation procedures of Connected and Automated Driving functions, including key technologies such as communications, cyber-security, and positioning. Tests were considered for different testing environments, from virtual simulations, Hardware-in-the-Loop (HiL), and proving ground testing to field-testing in the real-world, to validate safety and security performance according to the key users' needs. Not all scenarios are applicable or relevant for the different types of automated vehicles, having different functionalities and operational design domains (ODD). A method has been developed to select scenarios and generate test cases based on a description of the driving functions and the ODD.

Additional attributes related to the Key Enabling Technologies are added to the description of the test cases for appropriate testing of the impact of these technologies on the functionality and consequently on safety. Procedures have been developed to allocate test cases to the available test instances: virtual simulation testing, system-in-the-loop (XiL) based testing, Proving Ground testing and public road testing [25].

Key learnings from the HEADSTART project include:

- Scenario Databases are a key element for CCAM verification and validation but there is still a lack of harmonization between different databases.
- Need for more harmonization in virtual simulation.
- Lack of compatibility between physical tooling and simulation tooling.
- Role of standards is paramount in establishing common ground and providing technical guidance

Next steps and needs for further work that have been identified include:

- CCAM systems must prove to be reliable in every possible driving scenario, for which a strong safety argumentation is needed.
- Standardization is in infancy, and still need time to mature
- Synchronization is needed to establish a common practice as many standards in this field are under development or have been recently published.
- Instead of many individual solutions, a single concrete approach should be used in a universally agreed manner, able to deal with a wide variety of scenarios including their creation, editing and parametrization.
- Therefore, it is necessary to move to the next level of standardization, in the concrete specification and demonstration of a commonly accepted Safety Assurance Framework (SAF) for the safety validation of CCAM systems.

The SUNRISE project [26], which started in September 2022 and follows up on HEADSTART, aims at establishing a common safety assurance framework, interconnecting silos and making them collaborate in a harmonized way. It involves 22 partners.

The main goal is to develop and provide a harmonized and scalable CCAM Safety Assurance Framework that fulfils the needs of different automotive stakeholders, for a continuously evolving number of use cases and scenarios.

Figure 5 provides an overview of the approach used in SUNRISE. The Safety Assurance Framework constitutes the central element. It is based on methodologies and tools and a data framework, all supporting a large set of scenarios to be used to carry out the safety assessment of all types of Automated Vehicles, not limited to passenger cars.

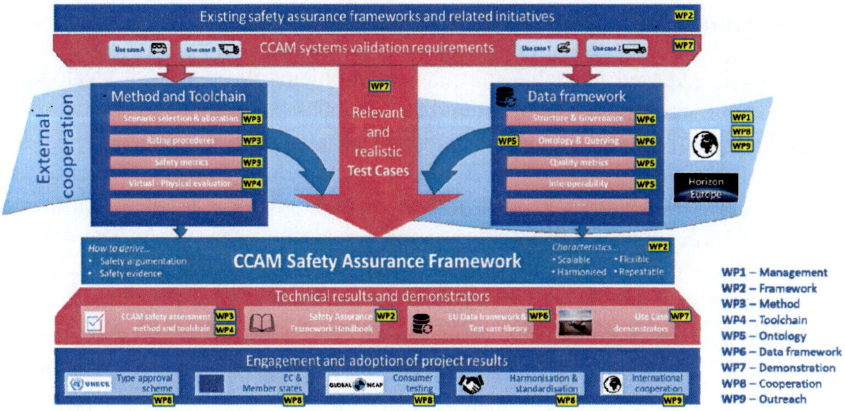

Fig. 5. Overview of the SUNRISE Approach and link with project Work Packages

Building on the results from HEADSTART, SUNRISE will improve methodologies for scenario-based verification and validation (V&V) testing and generate recommendations for harmonisation, standardisation and homologation entities. This will be done with the involvement of partners from the US, Japan, Canada, South Korea, Singapore, Australia, through a stakeholder network and cooperation platform. In particular, the project will work on:

- Widening the scope of use cases with a variety of environmental conditions, mixed traffic situations and edge cases
- Development of a Toolchain with tools allowing for their further development and adaptation with future technological evolution.
- Definition and development of methodologies and processes to continuously identify relevant events from various sources and convert them into detailed scenarios to support the development of a European Scenario Database. This will be complemented by an ontology.
- Standardised, open interfaces and quality controlled data exchange to enable the testing of multitude of relevant test cases.
- Seamless use of validated models from different sources.
- Scenario-based approaches combining virtual and physical testing.
- Conceptual description of a type approval scheme for CCAM systems considering all types of vehicles.
- Feed outcome into on-going discussions regarding EU type vehicle approval rules as well as in the framework of the UNECE.

The Framework to be developed, should allow for easy intergration of future ADS technologies and will be designed in such a way that future scenarios and parameters can be incorporated.

3 Conclusions

Collaboration is required to tackle the massive problems faced in the development, testing, and deployment of AV fleets around the world. Collaboration may be in the context of pre-competitive activities across companies, standards development, and testing tools that are open and interoperable. But efforts to develop standards, databases, and testing and safety frameworks are still in their early days. In the marketplace of ideas some tools and frameworks will likely see greater adoption than others. The ones that remain must also become interoperable with each other. Key words mentioned during the presentations included *collaboration*, *harmonization*, and *standardization.*

It has been suggested that a single concrete approach should be agreed upon to deal with a wide variety of scenarios. While this type of convergence may occur in countries and even some continents, it is unlikely that we will every reach universal agreement. Progress towards this end requires a number of factors, of which we note only two:

1. Existing methods, standards, and frameworks need to be continually exercised and stressed in competitive and cooperative contexts so that the best ideas can percolate up.
2. Conversion between standardized formats must exist at every level (e.g. scenario, road, ODD, network interface files, etc.).

For example, it should be well understood how to translate a left-driving scenario to a right-driving environment and vice-versa. Ideally, such a conversion ought to be contained to difference in roadway files with minimal changes needed to the scenario itself. This, and other tricky cases, make the end goal far from being easy or simple.

4 Next Steps

This year's session showed that good progress continues to be made in testing and data collection frameworks. We suggest the following list of action items:

- Since standards development is such an active area right now, effort should be made, either now or after upcoming standards have been published, to synchronize them and identify remaining gaps.
- Several safety measures should be integrated into testing frameworks so that commonly accepted metrics are easy to obtain from test suites. The various choices should continue to be compared and evaluated to match the most appropriate metrics, scenarios, and use cases (e.g. internal testing vs. certification vs. regulation).
- Develop databases and workflows that collect in-vehicle data and use it to generate test scenarios and even provide real-world sensor data.
- Additional rounds of collaborative projects should be funded to further exercise new testing frameworks with real-world data, scenarios and use cases.

References

1. Li, T., Han, X., Ma, J., Ramos, M., Lee, C.: Operational safety of automated and human driving in mixed traffic environments: a perspective of car-following behavior. Proc. Inst. Mech. Eng. Part O: J. Risk Reliab. 1748006X211050696 (2021)

2. Hu, X., Zheng, Z., Chen, D., Zhang, X., Sun, J.: Processing, assessing, and enhancing the waymo au- tonomous vehicle open dataset for driving behavior research. Transport. Res. Part C: Emerg. Technol. **134**, 103490 (2022)
3. DOT, U.: Next generation simulation (NGSIM) vehicle trajectories and supporting data (2016)
4. Geiger, A., Lenz, P., Stiller, C., Urtasun, R.: Vision meets robotics: the kitti dataset. Int. J. Robot. Res. **32**(11), 1231–1237 (2013)
5. Sun, P., et al.: Scalability in perception for autonomous driving: Waymo open dataset. In: Proceedings of the IEEE/CVF Conference on Computer Vision and Pattern Recognition, pp. 2446–2454 (2020)
6. Xu, R., Xiang, H., Xia, X., Han, X., Li, J., Ma, J.: Opv2v: an open benchmark dataset and fusion pipeline for perception with vehicle-to-vehicle communication. In: 2022 International Conference on Robotics and Automation (ICRA), pp. 2583–2589. IEEE (2022)
7. European Data Strategy (2019). https://commission.europa.eu/strategy-and-policy/priorities-2019-2024/europe-fit-digital-age/european-data-strategy_en. Accessed 22 Feb 2023
8. FOT-Net Data and CARTRE (2019), Data Sharing Framework, Version 1.1, 2019. https://www.connectedautomateddriving.eu/wp-content/uploads/2021/09/Data-Sharing-Framework-v1.1-final.pdf. Accessed 22 Feb 2023
9. FESTA (2021). FESTA Handbook, Version 8, 2021. https://www.connectedautomateddriving.eu/methodology/festa/. Accessed 22 Feb 2023
10. FOT-Net Data website. https://wiki.fot-net.eu/index.php?title=FOT-Net-Data. Accessed 22 Feb 2023
11. CARTRE Project on European Commission TRIMIS website. https://trimis.ec.europa.eu/project/coordination-automated-road-transport-deployment-europe. Accessed 22 Feb 2023
12. ARCADE (ConnectedAutomatedDriving) Project website. https://www.connectedautomateddriving.eu/about/arcade/. Accessed 22 Feb 2023
13. European General Data Protection Regulation (GDPR) website. https://gdpr.eu/. Accessed 22 Feb 2023
14. Kalisvaart, S., Pattinson, J.A., Svanberg, E., Barnard, Y, Hiller, J.: Proceedings of Workshop on Data Sharing (2021). https://knowledge-base.connectedautomateddriving.eu/wp-content/uploads/2021/03/EU-CSA-ARCADE_Proceedings_Workshop_Data_sharing_20210225.pdf. Accessed 17 Jan 2023
15. L3Pilot project website: L3pilot.eu, accessed 17 January 2023
16. Gaia-X. https://www.data-infrastructure.eu/GAIAX/Navigation/EN/Home/home.html. Accessed 23 Mar 2021
17. International data spaces association. https://www.internationaldataspaces.org. Accessed Mar 23rd 2021
18. PrepDSpace4Mobility Data ecosystems inventory. https://mobilitydataspace-csa.eu/inventory/?filter=map_location_taxonomy%3Droad-traffic. Accessed 22 Feb 2023
19. FAME Project (2023). https://www.connectedautomateddriving.eu/about/fame/. Accessed 22 Feb 2023
20. EU-wide Knowledge Base on CCAM. https://www.connectedautomateddriving.eu/. Accessed 22 Feb 2023
21. Data for Road Safety (DFRS) initiative website. https://www.dataforroadsafety.eu/. Accessed 22 Feb 2023
22. Vision Zero Network website: What is Vision Zero. https://visionzeronetwork.org/about/what-is-vision-zero/. Accessed 17 Jan 2023
23. Declaration of Amsterdam "Cooperation in the field of Connected and Automated Driving. https://voorlichting.rijksoverheid.nl/documenten/rapporten/2016/04/29/declaration-of-amsterdam-cooperation-in-the-field-of-connected-and-automated-driving
24. HEADSTART Project website. https://www.headstart-project.eu/. Accessed 22 Feb 2023

25. De Gelder, E., Op den Camp, O., A quantitative method to determine what collisions are reasonably foreseeable and preventable. Submitted to 8th Road Safety & Simulation International Conference, Athens, 8–10 June 2022
26. SUNRISE Project website. https://ccam-sunrise-project.eu/about/. Accessed 22 Feb 2023

Part IV: Transport System Planning

Inconsistency of AV Impacts on Traffic Flow: Predictions in Literature

Xiaowei (Tom) Shi[1(✉)], Hao Liu[2], Meng Wang[3], Xiaopeng Li[4], Biagio Ciuffo[5], Daniel Work[6], and David Kan[7]

[1] University of Michigan, 2350 Hayward Street, Ann Arbor, MI 48109, USA
tomcee@umich.edu
[2] University of California, Berkeley, 1357 S. 46th Street, Richmond, CA 94804, USA
liuhao@berkeley.edu
[3] Technische Universität Dresden, Hettnerstraße 1, 01069 Dresden, Germany
meng.wang@tu-dresden.de
[4] University of Wisconsin-Madison, 1415 Engineering Drive, Madison, WI 53706, USA
xli2485@wisc.edu
[5] European Commission Joint Research Centre, European Commission, Rue du Champ de Mars 21, 1050 Brussels, Belgium
Biagio.CIUFFO@ec.europa.eu
[6] Vanderbilt University, 1025 16th Avenue S, Suite 102, Nashville, TN 37212, USA
dan.work@vanderbilt.edu
[7] Florida Atlantic University, 777 Glades Road, ST 203, Boca Raton, FL 33431, USA
kanx@fau.edu

Abstract. In the literature, automated vehicle (AV) modeling studies tend to depict positive impacts of AV technologies on traffic. However, recent field experiments of production AVs (production vehicles with automated driving features) showed negative impacts on traffic flow stability and capacity. These inconsistencies may hinder the development and deployment of AV technologies. To identify major causes of the discrepancy, a breakout session was held at the 2022 Transportation Research Board (TRB) Automated Road Transportation Symposium (ARTS). Leading researchers from academia, industry, and government agencies were invited to present their thoughts on the issue. This book chapter summarizes the essence of the presentations and discussions at the breakout session. It provides insights into the modeling and simulation of AVs, AV technology development, and traffic management in the era of AVs.

Keywords: automated vehicles · traffic flow impacts · traffic simulation · field experiments · adaptive cruise control

1 Introduction

The past decade has witnessed rapid advancement in automated vehicle (AV) driving technologies such as adaptive cruise control (ACC), lane keeping, autonomous parking, and emergency braking. AV driving technologies have the potential to revolutionize the

G. Meyer and S. Beiker (Eds.): ARTSymposium 2022, LNMOB, pp. 165–173, 2023.
https://doi.org/10.1007/978-3-031-34757-3_13

future transportation system. Before large-scale deployment of any new technology, it is important to understand its benefits and costs. This helps society make appropriate decisions about how to utilize the technologies to maximize the benefits while minimizing the negative impacts.

The AV concept dates back to the late 1960s (Earnest 2012). It is not surprising that numerous studies have investigated the impacts of AVs on traffic (Shladover 2008). Many predicted that AVs could bring great benefits to our life, for example, improving road safety, enhancing traffic efficiency, and reducing fuel consumption and emissions. Nonetheless, the predicted benefits are often based on simulation experiments rather than real-world data. These simulation studies tend to envision an ideal AV operation environment, where road conditions and traffic patterns are well-controlled and predictable. While these simulations may suggest theoretical benefits of AVs, they cannot accurately consider the complexities and uncertainties of the real-world driving environment.

Thanks to the rapid development of AV technologies in recent years, a series of AV driving features are available in production vehicles. The availability of AVs enables us to collect empirical evidence and validate the theoretical analysis of AVs via field experiments. Those studies identified noticeable inconsistencies between the simulation and field experiments. For example, it was reported that the adaptive cruise control system is not string stable (Knoop et al. 2019; Gunter et al. 2019; Shang and Stern 2021; Shi et al. 2022). String unstable ACC vehicles can intensify small speed perturbations of the downstream vehicles, leading to serious traffic congestion to the upstream traffic. In addition, the existing ACC design may not improve the road capacity. The ACC system allows drivers to customize the following headway with their preceding vehicle. It was reported that with the longest following headway setting, the road capacity of AV traffic is even worse than that of human-driven vehicle traffic (Li et al. 2021; Shi and Li 2021a).

These inconsistent findings may hinder the further development and deployment of AV technologies. To identify major reasons causing this discrepancy and develop remedies, a breakout session was held at the 2022 Transportation Research Board (TRB) Automated Road Transportation Symposium (ARTS). Leading researchers from academia, industry, and government agencies were invited to present their thoughts to tackle the inconsistent issue. This book chapter provides insights into the modeling and simulation of AVs, AV technology development, and traffic management in the era of AVs based on the presentations and discussions of the breakout session. Without causing confusion, the AVs discussed in this book chapter represent a broad concept e.g., connected and autonomous vehicles (CAVs), electric AVs, etc.

2 Literature Review and Session Presentation Summary

This section first reviews the simulation studies on the impact of AVs. Then, each speaker's presentation summary is provided, highlighting the inconsistent results obtained from field experiments compared to the simulation experiments. Their suggestions and ongoing research efforts were also presented. Finally, the section concludes by showcasing the advanced modeling and simulation tools that may reduce the inconsistency.

2.1 AV Impacts with Simulation Experiments

Microscopic traffic simulation is a primary tool for assessing the impacts of vehicle-highway automation on traffic performance of freeways and arterial corridors. By analyzing the simulation results, researchers can gain insights into how AV impacts traffic flow, safety, and energy efficiency. This can help inform policy decisions and improve the design of transportation systems.

There are two major approaches to build microscopic models required by the simulation experiments. The first approach recalibrates the parameters of existing models based on data sets from field tests of vehicle-highway automation applications. This work allows the existing models to depict the behaviors of both manually driven vehicles and AVs under the influence of advanced traffic management strategies. Since this approach takes advantage of the existing model structure and calibration methods, it does not require high model development effort. However, as the existing models are originally built for human drivers, they may lack the capability to capture the unique behaviors of AVs, such as vehicle-to-vehicle and vehicle-to-infrastructure cooperation and trajectory optimization. Examples of this modeling approach can be found in Calvert et al. (2017) and Kesting et al. (2007).

The second modeling approach creates specific models for AVs based on the field data. This approach includes a selection of the best model structure that minimizes the differences between the model output and the field observations. It also provides the flexibility to keep the control logic of the AVs in the resulting traffic models. This modeling technique requires a detailed understanding of vehicle dynamics control and system identification. Thus, the modeling effort is expected to be high. Examples of this approach can be found in Milanes et al. (2014) and Xiao et al. (2017).

Models resulting from both approaches face challenges of model transferability. Usually, researchers develop a model based on field data that covers limited traffic and vehicle operating conditions. But they may need to implement the model in analysis scenarios that are not represented by the field data. Such a practice could raise uncertainties regarding the model outputs. It is unclear to what degree the model outputs match the true system behaviors. Reducing the output uncertainties requires model evaluation with different data sets in the model development stage. Another limitation of existing models is caused by the lack of human factor components. Before vehicles are fully automated, human drivers and automated controllers will operate the vehicle in parallel. The transition of control (ToC) between the driver and the controller could substantially affect the behavior of AVs and CAVs in the traffic stream. The development of ToC functions is critical for building a high-fidelity microscopic model.

Given the limitations associated with simulation experiments, studies that solely rely on the simulation experiments to investigate the impact of AVs on traffic are subject to inaccuracies. For example, Karaaslan et al. (1991) predicted that AVs could increase the freeway capacity by a factor of four. This result seems a little bit too optimistic. In recent studies, researchers claimed that AV might achieve a double or triple improvement in road capacity (Lioris et al. 2017; Olia et al. 2018). However, some researchers found that considering the current conservative legislation, AVs may even have a negative impact on the capacity of freeways (Hartmann et al. 2017; Kakimoto et al. 2018). To this end, to produce more accurate and reliable results, studies that employ field experiments to validate the relevant AV impacts findings in the literature are necessary.

2.2 Inconsistent Findings with Field Experiments

As AV driving features are increasingly equipped on production vehicles, a few pioneer researchers investigated the impacts of AV by conducting field experiments. We invited four leading researchers in this field from academia to present their recent works. Their presentations and main outcomes are summarized as follows.

2.2.1 Traffic Flow Smoothing with Level 1 Automated Vehicles

Daniel Work, Vanderbilt University

Phantom jams, also known as "spontaneous traffic jams," are congestion that occurs without any apparent cause, such as accidents or roadblocks. Field experiments (Sugiyama et al. 2008) showed that these jams can occur due to human driving behavior alone. Since these phantom jams also exist in real traffic, is it able to control a few level-1 AVs in traffic to make the overall traffic smoother? Dr. Work gave his answer to this question in his presentation.

According to the field experiments conducted by Dr. Work's group, it was found that the commercially implemented ACC system is not string-stable (Gunter et al. 2020). It means that if these production AVs are deployed into traffic, these vehicles cannot smooth traffic. Thus, Dr. Work proposed a series of new vehicle controller using information about the traffic ahead. By deploying this new controller to a few real-world AVs and testing at a segment of I-24, it was observed that: without enabling the controller, more stop-and-go patterns happened, and more fuel was used for all vehicles; with enabling the controller, the traffic flow became more uniform, and less fuel was used (Lichtle et al. 2022). It followed earlier experimental work that demonstrated the ability of controllers to smooth jams and reduce fuel consumption (Stern et al. 2018) (Stern et al. 2018). These observations answer the question raised previously and provide valuable insights into the potential of low-level automation to improve existing transportation systems.

Dr. Work mentioned that a larger-scale experiment would be conducted on the I-24 in Fall 2022. His team is working on the installation of 4K resolution video cameras and modern computer vision algorithms on the segment of I-24, which will provide four miles of continuous camera coverage observing all vehicles (Gloudemans et al. 2023). Thus, it is expected to better support the proposed controller that needs the input of the other vehicles' trajectories in traffic. The trajectory data obtained from the high-resolution videos will be publicly available.

2.2.2 From Automated Control to Cooperative Driving Automation

Xiaopeng Li, University of Wisconsin-Madison

Cooperative Driving Automation (CDA) offers numerous benefits to future transportation, such as improving road safety, enhancing traffic efficiency, and reducing fuel consumption and emissions. Furthermore, CDA can provide increased comfort for passengers by reducing the need for them to be actively involved in driving and enabling them to relax or attend to other tasks. These benefits demonstrate the potential of CDA to revolutionize the way we travel, making our roads safer, more efficient, and more enjoyable. To achieve these benefits of CDA, complicated coordination of vehicles, infrastructures, and control units is required. Specifically, from the vehicle side, it needs precise control of the AV trajectory.

According to the results of field experiments with production AVs, it is found that there is a significant gap between target and actual vehicle trajectories. Also, speed profiles are inconsistent across different runs. These issues may hinder the future development and deployment of CDA.

One solution proposed by Dr. Li is to incorporate a safety buffer into AV control. Dr. Li proposed a reactive AV car following control model, which reveals the trade-off between safety, mobility, and stability in AV following control (Li 2022). Field experiments of production AVs were conducted to verify the trade-off relationships (Shi and Li 2021b). The other solution proposed by Dr. Li is to incorporate the future trajectory of the AV into its control. Dr. Li proposed an online reinforcement learning-based model predictive control (MPC), which can accurately control a vehicle's speed under varying environments. Robot cars were used to test the performance of the proposed method.

Based on Dr. Li's presentation, the main takeaways are 1) Existing production AV control may not be ideal for complicated CDA trajectory control; 2) Advanced control methods (e.g., integrating learning, MPC, etc.) need to be integrated into production AV to enhance control performance.

2.2.3 Modeling Traffic Flow of AVs, and What if They are also Electric Vehicles (EVs)?

David Kan, Florida Atlantic University

As pointed out by Dr. Kan, AV traffic modeling and simulation performance largely relies on the data to build the model. The inconsistent impact of AVs on traffic flow between simulation and field experiments is primarily due to the lack of standardized AV data collection procedures. To accurately assess the impacts of AV on traffic flow, it is crucial that data collection processes capture a full range of speeds of the AVs.

To this end, Dr. Kan developed a new protocol for AV data collection (Kan et al. 2022). Production AVs equipped with high-definition GPS and OBD II data logger were used to collect the vehicle data. The experiment speeds varied from 0 to 60 mph. The experiments were conducted at off-peak hours on remote public roads without interruption from other road users. With the collected production AV data, Dr. Kan developed a microscopic-level car-following model and used it to simulate macroscopic-level performance. Based on the results, Dr. Kan reported that the existing production AVs would cause a lower discharge flow rate and higher delay.

Dr. Kan pointed out that the internal combustion engine may be a reason causing this result. Compared to the internal combustion engine, the electric motor has higher tolerances for acceleration and braking and less delay due to direct drive. Thus, if the production AV is combined with the electric motor, the vehicles are expected to have better performance than the production AV with the internal combustion engine.

2.2.4 Empirical Study on the Properties of Adaptive Cruise Control Systems and Their Impact on Traffic Flow and String Stability

Ciuffo Biagio, European Commission Joint Research Centre

AVs promise to significantly improve road traffic. To a certain extent, this situation is similar to the expectations at the end of the last century about the positive effects that the introduction of ACC systems would have had on motorway traffic. The parallelism is interesting because ACC-equipped vehicles represent the first level of vehicle automation and are now widely available on the market. In this light, studying ACC impacts can help to anticipate potential problems related to its widespread application and to avoid that AVs will lead to the same problems.

In Dr. Ciuffo's presentation, he introduced the results of a large-scale experimental campaign involving ACC vehicles, which has allowed us to quantify the impacts of AV on motorway driving and compare them with the original expectations and the underlying assumptions made to justify them (Ciuffo et al. 2021; Makridis et al. 2021). His presentation concludes with possible recommendations to avoid what we see today for ACC vehicles that will not happen in 10–20 years for AVs.

2.3 Advanced Modeling and Simulation Tools for AVs

This section summarizes the advanced modeling and simulation tools utilized by industry companies and government agencies to understand the impact of AVs on traffic.

2.3.1 Data Collection to Improve Simulation of Connected and Automated Vehicle (CAV) Driving Behavior

Transportation agencies require a cost-effective solution for determining the impact of CAV deployments to make intelligent investment and operational decisions. Traffic analysis tools offer an efficient means of evaluating new technologies and strategies prior to implementation. However, existing Advanced Mobility Simulation (AMS) tools were designed to model human driving behavior. New CAV behavioral models are required to be integrated into microsimulation tools.

The speaker introduced the CAV AMS research program of FHWA, which is composed of three aspects: data collection, model improvements, and benefits estimation. The objective of the FHWA data collection projects is to collect robust datasets about the behavior of CAVs and surrounding traffic in naturalistic conditions. The three ongoing projects of FHWA are CARMA Data Collection, Acquiring CAV Performance Datasets, and Third-Generation Simulation: A Closer Look at the Impacts of Automated Driving Systems (ADS) on Human Behavior. The datasets will be used to improve traffic simulation models and will be made available under Creative Commons Zero (https://creativecommons.org/share-your-work/public-domain/cc0/) and posted on Transportation.gov. Three recent publications of FHWA were also introduced by the speaker. They are the CAV AMS cornerstone framework, CAV model improvement: tools development, and CAV benefits estimation: case studies.

The speaker's presentation pointed out that 1) Aerial data collection can be used to collect vehicle trajectories for non-instrumented vehicles efficiently; 2) Vehicle sensor data can be used to mine vehicle trajectories for model development and calibration in the future; 3) Ground truth data improve our ability to simulate how CAVs' behavior will impact system performance.

2.3.2 Are Traffic Simulators the Right Tool for Evaluating Connected and Autonomous Vehicles?

Traffic simulation models are computer-based representations of real-world traffic systems. They aim to simulate and analyze various aspects of traffic, such as traffic flow, congestion, and network capacity, to make predictions about how the traffic system will behave under different conditions. The models use mathematical algorithms and statistical data to simulate the interactions between vehicles, road networks, traffic signals, and other components of the traffic system.

To incorporate connected and/or automated vehicles into traffic simulation models, the behaviors of the vehicles need to be well studied. The speaker introduced the way that Aimsun (https://www.aimsun.com/) studies AV/CAV behaviors. The Aimsun simulation platform models AV behaviors from four aspects: car-following, lane-changing, gap acceptance, and cooperation. To develop and test CV applications, the Aimsun simulation platform integrates a V2X communication API, which introduces the concept of packet latency and packet loss and creates VANets between connected vehicles and roadside devices within a maximum transmission range. Also, the V2X sensors can be configured in the network model. Thus, microscopic simulations between vehicles and infrastructure can be performed.

With this, the impacts of AV can be studied with the Aimsun simulation platform from the following aspects. 1) Microscopic simulation-based experiments to derive the network capacities through the network Macroscopic Fundamental Diagram (MFD); 2) Statistical analysis for the identification of the effects on the PCUs. Estimating PCU functional relationship; 3) Using PCU functional relationship as input to the VDFs of macroscopic demand models to forecast impacts on network performance. Inform policy-making for optimal AV share.

The speaker emphasized that the real-world CAV dataset is critical for developing traffic simulation models. Due to the lack of data, uncertainty should be considered when we design the CAV models. Also, the speaker pointed out that traffic simulations are not a perfect tool to model CAVs but are necessary, which answers the question in his presentation title.

3 Conclusions and Next Steps

In the existing literature, AV modeling studies without empirical evidence/validation tend to support the positive impacts of AV technologies on traffic. However, recent field experiments of production AVs showed negative impacts on traffic flow stability and capacity. This inconsistency may hinder the further development and deployment of AV technologies. This book chapter introduces the major reasons causing this discrepancy and the way to close the gap. The conclusions of this book chapter and the suggested future work are as follows:

1) Multiple methods are available to mitigate ACC's negative impacts on traffic flow, including
   - Faster sensing and perception technologies for reducing the system reaction time
   - Advanced control algorithms that provide stable following behaviors
   - New powertrain options such as EVs that enable efficient power output
   - Regulations that promote ACC designs that benefit the traffic system
2) New AV/CAV datasets are/will be available to the public
   - Need reliable tools to generate accurate information from the data
   - Need to streamline data processing, information extraction, AV/CAV system evaluation, and traffic model building
   - Opportunities to design creative ways to collect AV/CAV data without high costs
3) Traffic simulation models are a viable tool to study AV/CAV, but require refinement
   - Need to reflect new AV/CAV technologies and human-machine interactions
   - Need to be scalable for road networks
   - Need to produce consistent results comparable to benchmark data sets

References

Calvert, S.C., Schakel, W.J., van Lint, J.W.C.: Will automated vehicles negatively impact traffic flow? J. Adv. Transp. **2017** (2017). https://doi.org/10.1155/2017/3082781

Ciuffo, B., et al.: Requiem on the positive effects of commercial adaptive cruise control on motorway traffic and recommendations for future automated driving systems. Transp. Res. Part C Emerg. Technol. (2021). https://doi.org/10.1016/j.trc.2021.103305

Earnest, L.: Stanford cart (2012). https://web.stanford.edu/~learnest/sail/oldcart.html

Gloudemans, D., Wang, Y., Ji, J., Zachar, G., Barbour, W., Work, D.B.: I-24 MOTION: An instrument for freeway traffic science. Electrical Engineering and Systems Science (2023). https://arxiv.org/abs/2301.11198

Gunter, G., et al.: Are commercially implemented adaptive cruise control systems string stable? IEEE Trans. Intell. Transp. Syst. **19122**, 1–12 (2020). https://doi.org/10.1109/TITS.2020.3000682

Gunter, G., Janssen, C., Barbour, W., Stern, R.E., Work, D.B.: Model based string stability of adaptive cruise control systems using field data. IEEE Trans. Intell. Veh. **5**, 90–99 (2019). https://doi.org/10.1109/TIV.2019.2955368

Hartmann, M., Motamedidehkordi, N., Krause, S., Hoffmann, S., Vortisch, P.: Impact of automated vehicles on capacity of the German freeway network. In: ITS World Congress, 2017Montr (2017)

Kakimoto, Y., Iryo-Asano, M., Orhan, E., Nakamura, H.: A study on the impact of AV-HDV mixed traffic on flow dynamics of single-lane motorway. Transp. Res. Procedia (2018). https://doi.org/10.1016/j.trpro.2018.11.035

Kan, P.C., Imran, M.A., Murshed, M.T., X.K.: Field Experiment on the Impact of Automated Vehicles on Arterial Capacity – Case Study of Adaptive Cruise Control (ACC) (2022)

Knoop, V.L., Wang, M., Wilmink, I., Hoedemaeker, D.M., Maaskant, M., Van der Meer, E.-J.: Platoon of SAE level-2 automated vehicles on public roads: setup, traffic interactions, and stability. Transp. Res. Rec. J. Transp. Res. Board **2673**, 311–322 (2019). https://doi.org/10.1177/0361198119845885

Karaaslan, U., Varaiya, P., Walrand, J.: Two proposals to improve freeway traffic flow. Proc. Am. Control Conf. **3**, 2539–2544 (1991). https://doi.org/10.23919/acc.1991.4791860

Kesting, A., Treiber, M., Helbing, D.: General lane-changing model MOBIL for car-following models. Transp. Res. Rec. J. Transp. Res. Board **1999**, 86–94 (2007). https://doi.org/10.3141/1999-10

Li, T., Chen, D., Zhou, H., Laval, J., Xie, Y.: Car-following behavior characteristics of adaptive cruise control vehicles based on empirical experiments. Transp. Res. Part B **147**, 67–91 (2021). https://doi.org/10.1016/j.trb.2021.03.003

Li, X.: Trade-off between safety, mobility and stability in automated vehicle following control: an analytical method. Transp. Res. Part B Methodol. **166**, 1–18 (2022). https://doi.org/10.1016/j.trb.2022.09.003

Lichtle, N., Vinitsky, E., Nice, M., Seibold, B., Work, D., Bayen, A.M.: Deploying traffic smoothing cruise controllers learned from trajectory data. In: 2022 International Conference on Robotics and Automation (ICRA), pp. 2884–2890. IEEE (2022). https://doi.org/10.1109/ICRA46639.2022.9811912

Lioris, J., Pedarsani, R., Tascikaraoglu, F.Y., Varaiya, P.: Platoons of connected vehicles can double throughput in urban roads. Transp. Res. Part C Emerg. Technol. **77**, 292–305 (2017). https://doi.org/10.1016/j.trc.2017.01.023

Makridis, M., Mattas, K., Anesiadou, A., Ciuffo, B.: OpenACC. An open database of car-following experiments to study the properties of commercial ACC systems. Transp. Res. Part C Emerg. Technol. (2021). https://doi.org/10.1016/j.trc.2021.103047

Milanes, V., Shladover, S.E., Spring, J., Nowakowski, C., Kawazoe, H., Nakamura, M.: Cooperative adaptive cruise control in real traffic situations. IEEE Trans. Intell. Transp. Syst. **15**, 296–305 (2014). https://doi.org/10.1109/TITS.2013.2278494

Olia, A., Razavi, S., Abdulhai, B., Abdelgawad, H.: Traffic capacity implications of automated vehicles mixed with regular vehicles. J. Intell. Transp. Syst. Technol. Planning, Oper. **22**, 244–262 (2018). https://doi.org/10.1080/15472450.2017.1404680

Shang, M., Stern, R.E.: Impacts of commercially available adaptive cruise control vehicles on highway stability and throughput. Transp. Res. Part C Emerg. Technol. **122**, 102897 (2021). https://doi.org/10.1016/j.trc.2020.102897

Shi, X., Li, X.: Constructing a fundamental diagram for traffic flow with automated vehicles: methodology and demonstration. Transp. Res. Part B Methodol. **150**, 279–292 (2021). https://doi.org/10.1016/j.trb.2021.06.011

Shi, X., Li, X.: Empirical study on car-following characteristics of commercial automated vehicles with different headway settings. Transp. Res. Part C Emerg. Technol. **128**, 103134 (2021). https://doi.org/10.1016/j.trc.2021.103134

Shi, X., Yao, H., Liang, Z., Li, X.: An empirical study on fuel consumption of commercial automated vehicles. Transp. Res. Part D Transp. Environ. **106**, 103253 (2022). https://doi.org/10.1016/j.trd.2022.103253

Sugiyama, Y., et al.: Traffic jams without bottlenecks-experimental evidence for the physical mechanism of the formation of a jam. New J. Phys. (2008). https://doi.org/10.1088/1367-2630/10/3/033001

Shladover, S.E.: AHS research at the California PATH program and future AHS research needs. In: Proceedings of 2008 IEEE International Conference on Vehicle Electronics Safety, ICVES 2008 (2008). https://doi.org/10.1109/ICVES.2008.4640915

Stern, R.E., et al.: Dissipation of stop-and-go waves via control of autonomous vehicles: field experiments. Transp. Res. Part C Emerg. Technol. **89**, 205–221 (2018). https://doi.org/10.1016/j.trc.2018.02.005

Xiao, L., Wang, M., Van Arem, B.: Realistic car-following models for microscopic simulation of adaptive and cooperative adaptive cruise control vehicles. Transp. Res. Rec. **2623**, 1–9 (2017). https://doi.org/10.3141/2623-01

Interactive Traffic Management for Highly Automated Vehicles

Tom Alkim[1](✉), Claudio Roncoli[2], Torsten Geissler[3], Siddartha Khastgir[4], and Risto Kulmala[5]

[1] MAP Traffic Management, Utrecht, The Netherlands
tom.alkim@maptm.nl
[2] Aalto University, Espoo, Finland
claudio.roncoli@aalto.fi
[3] Federal Highway Research Institute (BASt), Bergisch Gladbach, Germany
geissler@bast.de
[4] University of Warwick, Coventry, UK
s.khastgir.1@warwick.ac.uk
[5] Traficon Ltd., Espoo, Finland
risto.kulmala@traficon.fi

Abstract. This chapter summarizes part of the presentations and discussions that took place at the Automated Road Transportation Symposium 2022 (ARTS22) during the breakout session titled "Interactive Traffic Management for Highly Automated Vehicles", complemented with further material and reflections by the authors.

The chapter deals with traffic management in the presence of highly automated vehicles capable of driverless operation in specific Operational Design Domains (ODD). The aim is to present and discuss challenges related to the design and implementation of traffic management measures from multiple perspectives, covering the exchanged messages, the digitization of traffic codes and regulations, the impact on efficiency, and the human factors and reactions to traffic management measures. The potential to increase and optimize the performance of highly automated vehicles by providing external information will also be explored by introducing the concept of Distributed ODD attribute Value Awareness (DOVA).

Keywords: traffic management · automated vehicles · connected vehicles · operational Design Domain

1 Introduction: Motivation and Objectives

An increasing presence of highly automated vehicles or Connected and Automated Vehicles (CAVs) is expected to have a significant impact on traffic management, while stakeholders, such as public agency operators and toll road operators, need to adapt their Traffic Management Center (TMC) policies and procedures. In the meantime, vehicle manufacturers need to adapt their in-vehicle systems to be able to react appropriately to traffic management messages.

G. Meyer and S. Beiker (Eds.): ARTSymposium 2022, LNMOB, pp. 174–182, 2023.
https://doi.org/10.1007/978-3-031-34757-3_14

The main objective of this chapter is to present and discuss recent developments and future challenges related to the development of traffic management for automated vehicles, looking from a multifaceted perspective. More detailed objectives include the following.

- To provide insights on the needs to develop traffic management of the future, including the roles and responsibilities of the stakeholders.
- To increase understanding of the digital, physical, and operational infrastructure associated with CAVs and traffic management.
- To introduce the concept of Distributed ODD attribute Value Awareness (DOVA).
- To discuss the roles and responsibilities of various stakeholders in translating traffic codes and rules into a machine-readable format, as well as the expected responses of CAVs to these codes and rules.
- To raise awareness on the impacts that operating CAVs in different infrastructure situations may have and the need for traffic management as a means to mitigate negative externalities.

2 Summary of the Discussion

2.1 CCAM (Meta) Taxonomies

Support from the physical and digital road infrastructure can extend the conditions under which connected and automated vehicles can operate safely. Operational Design Domain (ODD) and Infrastructure Support for Automated Driving (ISAD) are key terms in taxonomies related to Connected Cooperative and Automated Mobility (CCAM) but do not yet provide the full picture that has emerged in recent years. It is important to take the full picture into account for collaborating between the actors across sectors (e.g., automotive industry, road infrastructure managers) in order to prepare, pilot, test, and deploy CCAM services in the coming decades, ultimately for the benefit of end users.

CCAM taxonomies comprise the automotive side standardized classifications, i.e., the Levels of Driving Automation (SAE J 3016) and the Cooperation Classes (SAE J 3216). They are complemented by taxonomies on infrastructure suitability such as Infrastructure Support for Automated Driving (ISAD) and Levels of Service for Automated Driving (LOSAD). In addition, they comprise the (automation mode) communication towards users, in order to provide mode-awareness and avoid mode-confusion. Figure 1 illustrates the CCAM taxonomies. The interplay between the taxonomies is analyzed in more detail in [1].

From an overall perspective, the taxonomies fit well with each other and come to quite consistent results. In a further step, it may be useful to investigate the possibilities of integration into a meta-taxonomy. This advancement should not be an end in itself, but always only a serving tool enabling us to achieve common goals. A recent and even more promising approach, the concept of Distributed ODD attribute Value Awareness (DOVA), is featured in the next subchapter. The cross-sector collaboration is a constitutional feature of it. In other words, what the meta-taxonomy would aim to tie together, is already built into the concept of DOVA.

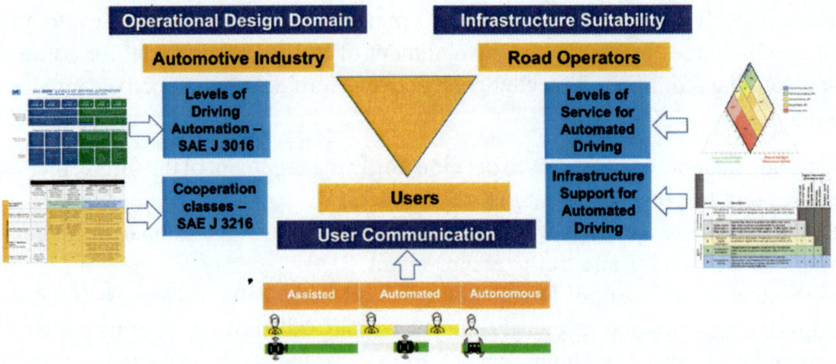

Fig. 1. CCAM Taxonomies by Geissler and Shi [1].

2.2 Concept of Distributed ODD Attribute Value Awareness

The need to monitor or be aware of each ODD attribute puts an additional overhead on CAVs or Automated Driving Systems (ADS) to be able to measure each ODD attribute. However, measuring each ODD attribute may not be practically feasible from a cost and engineering perspective. However, ODD awareness is key to ensuring the safe operation of the ADS. In order to overcome this challenge, we introduce the concept of Distributed ODD attribute Value Awareness (DOVA) framework building on the DOA framework introduced by Khastgir et al. [2].

The DOVA framework enables the ADS to benefit from off-board sensing infrastructure to become aware of ODD attribute values that may not be able to be measured or sensed by themselves. For example, an ADS may not be able to detect the severity of a visibility impairment from a fog bank that it is approaching. It may be able to receive such information from a roadside weather station that can provide this information through over-the-air communication with the ADS. This enables the ADS to have awareness of this current operating condition and compare it with its designed ODD to establish if the ADS is either inside or outside its ODD.

While the information for some of the ODD attributes could be available via infrastructure, there may potentially be commercial services that can augment ODD information for the ADS.

From a road operator's or traffic manager's perspective, it is important to establish what type of ODD attribute information should be provided via infrastructure and its corresponding quality to enable the safe deployment of ADS. It is also important to consider the needs of road operators and traffic managers to be aware of any ADS approaching the end of their ODD and/or being in a transitional or minimal-risk state.

The operation of the DOVA framework in practice is illustrated in Fig. 2. The ODD attribute information (or from the road operator perspective, local condition attribute information) sharing plays a major role in influencing the driving behavior of a CAV, depending on its technical capabilities and the rules of the road. The traffic management operations affect the rules of the road (i.e., the expected behavior) as well as the status of the ODD / local condition attributes sensed by the vehicle, the road operators' and other stakeholders' monitoring and other data acquisition systems providing the attribute information to the ADS-operated vehicles and other road users.

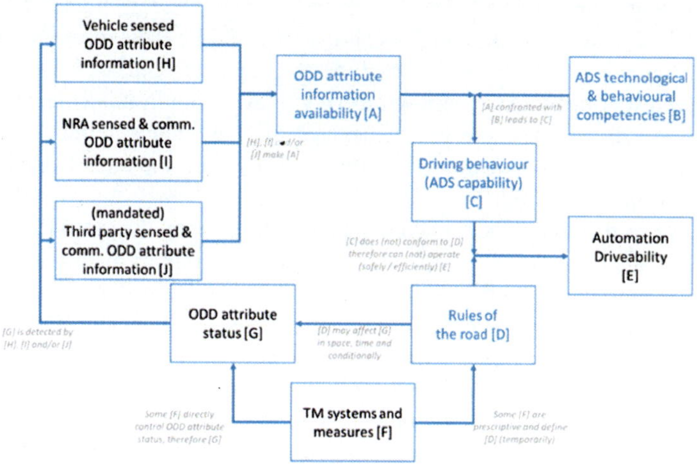

Fig. 2. Distributed ODD Awareness Framework by Khastgir et al. [2].

2.3 Impact of CAVs on Traffic Flow

Traffic congestion in and around urban areas has a serious detrimental impact on the economic and social life of modern society, as well as on the environment, with negative effects for climate change, which calls for radical solutions [3]. Insofar, typical responses to mitigate traffic congestion externalities have been the expansion of road infrastructure, with huge costs and impact on the environment, and the implementation of traffic management measures, which however may also face various kinds of limitations. The appearance of CAVs has come with the promise to drastically improve safety and mobility, essentially revolutionizing the way how people move around. Whether CAVs have the potential to actually deliver the full range of expected benefits will ultimately depend on three factors: their penetration speed, their effectiveness, and their potential negative impacts. Often, studies tend to be overly optimistic about the future of CAVs by overestimating the first two factors while ignoring the third one. One reason is that vehicles are products manufactured and offered in a market, wherein there is competition for the customers' preferences; thus, it is reasonable to expect that, if no interventions are made, CAVs (are and) will be developed to benefit the individual vehicle and driver, often without a clear view or understanding for the implications, including potential advantages and disadvantages, that they may have for the induced, accordingly modified traffic flow characteristics.

One important aspect to consider is that CAVs are going to be *safe by design*, meaning that it is expected that no vehicle released on a large scale in the market is going to be taking any risk while driving, for either the occupants or other users surrounding the vehicle. However, this is not what typically happens with a vehicle driven by humans, who, actually, and in particular while traffic flow is perceived high, typically take risks, such as performing cut-in maneuvers or driving with (too) low headway distances from the preceding vehicle. Most often these risks do not lead to serious consequences, and the reason is that human drivers have developed a good perception and capability of anticipating the behavior of other drivers and what could be their reactions caused by such risky maneuvers. Also, other external factors, together with formal and informal communication that takes place between drivers, play a role in this context (e.g., eye contact between drivers, and the use of devices such as indicator lights and horns). However, all these aspects are extremely difficult to be coded and included in the design of automated vehicle operation.

The effects of the above-mentioned combined factors may thus lead to a deterioration of traffic characteristics, such as road capacity. This implies that an infrastructure designed to accommodate a certain capacity may not be able to sustain the same number of vehicles, which, in turn, would deteriorate the efficiency of the current transportation system. Various research has tried to quantify such reduction, resulting in numerical differences but similar trends [3–5]. The most promising solution to such an issue involves moving forward with the design and deployment of cooperative connected and automated systems, where automated vehicles should not behave as selfish entities, but they should act in a coordinated manner, either through centralized or decentralized management policies to achieve a benefit for the overall traffic system. Several management strategies have been proposed to improve the efficiency of systems with connected and automated vehicles, also in the presence of mixed traffic (see, e.g., [6–8]), which have the potential to mitigate the externalities potentially caused by automated vehicles on traffic characteristics, bringing unprecedented benefits for the overall traffic system performance. In fact, connected and automated vehicle systems provide increased opportunities for more efficient implementation of management strategies, where CAVs may actuate more precise commands ordered by, e.g., infrastructure-based intelligence. Furthermore, the granularity of traffic management measures, which currently depends on the installation and operation of dedicated infrastructure, could be arbitrarily defined once it is enabled by automated vehicles, and there could be different measures implemented, e.g., by lane, by destination, or by other criteria for vehicle selection.

2.4 Codifying Traffic Rules for CAVs

While driving, human drivers are expected to adhere to road traffic rules to ensure a smooth and safe flow of traffic on roads. Similarly, CAVs will also be expected to follow traffic rules to ensure a smooth flow in heterogenous traffic as well as ensure predictability of their behavior. Thus, as part of the assurance process for CAVs, they will need to be tested against the road rules. At the same time, an NRA would like to ensure that CAVs driven on their road networks also adhere to road traffic rules.

Rules of the road for human-driven vehicles require a certain level of interpretation on the part of the human drivers where they make an intuitive driving decision (e.g., understanding when is there a “suitable gap” to overtake a vehicle on the road). Such judgment will need to be made and hardcoded into the intelligence systems of the CAVs. In order to verify their behavior’s compliance with the rules of the road, there is a need to codify the rules of the road to enable objective assessment.

To this end, an ODD-based codification process for the rules of the road has been proposed by the United Nations Economic Commission for Europe (UNECE) Functional Requirements for Automated Vehicles (FRAV) informal working group [9]. The approach is also relevant for NRAs to understand the relevance of specific traffic rules and the CAVs' driving decisions when deployed in a specific ODD.

The operational design domain (ODD) refers to the operating environment in which vehicles can operate safely. As defined in the BSI PAS 1883 ODD taxonomy [10], it covers environmental conditions such as rainfall, scenery elements such as drivable areas, and dynamic elements such as macroscopic traffic behavior and designated speed of the subject vehicle.

If one compares the scope of ODD and the content of current "rules of the road for human drivers" (e.g., the UK's Highway Code or the Vienna Convention's Rules of the Road), a large overlap of scenery aspects and environmental conditions aspects can be observed.

Any rule of the road can be classified into two categories:

- *Doing* some "behavior" "somewhere"
- *Not doing* some "behavior" "somewhere"

While doing or not doing some behavior can be defined as part of CAVs' behavior capabilities, "somewhere" could be considered as an "operating condition" or part of the ODD definition. Thus, each rule of the road is a combination of aspects of ODD and behavior capabilities.

From an NRA perspective, understanding the ODD definition, i.e., the various ODD attribute values and their relationship with real-world routes will provide guidance on the applicable rules of the road for the CAVs. This can enable traffic flow planning and monitoring in a heterogenous traffic flow.

2.5 Remote Monitoring and Support for CAVs

For most of the current deployments of CAVs around the world, remote monitoring and/or support is being used and it is regarded as a necessity. Perhaps over time, when CAV's capabilities increase, such support may be minimized or even become obsolete, but for the foreseeable future it is certainly indispensable.

Remote monitoring and support involve a human operator providing instructions, permissions, or waypoints to the vehicle, or remotely driving it when the vehicle cannot execute one or more of its driving tasks [11]. It is considered most useful when the vehicle encounters unknown situations or when illegal actions are required. Remote support has many benefits, such as enabling operations, ensuring safety, and increasing public acceptance. The purpose and tasks of the operator can be very diverse for different modes and environments, ranging from confined areas for cargo movements to passenger vehicles on public roads. There are four levels of remote support: no assist, remote assist, remote control, and shared control [11]. Remote control is temporary full operational control typically used to resolve a situation, while shared control involves remote human driving with the vehicle controlling the on-board crash avoidance systems or remote assessment of a situation and providing concrete operational guidance recommendations.

The key elements of remote support for CAVs are a stepwise approach to building experience and trust with such operations, addressing human factors in remote operations, and defining the role of the human operator. There is still a need for further research on higher operational speeds and resulting increased safety risks, investigating edge cases, and looking at the ODD from a system-to-system perspective beyond the scope of only the vehicle.

Overall, remote support is expected to play a crucial role in achieving safe and comfortable highly automated transport services in mixed traffic. The implementation of remote support requires addressing various technical and operational challenges, and further research is needed to ensure its effective deployment.

2.6 From Automating Vehicles to Automating Traffic

Up until now, the design of AVs and CAVs by manufacturers has considered essentially a static road network, where infrastructure characteristics and regulations are embedded in the vehicle system and do not respond to dynamic external inputs [12, 13]. The dynamic components have been limited to interactions with other vehicles or obstacles, essentially only for safety purposes, while decisions are taken advantaging the individual vehicle only. Limited work has been done on standardizing messages for Advanced Traveler Information System [14], whereas there is no standardization on the response that vehicles may implement once such messages are received. On the other hand, a large body of research (see, e.g., [6]) has suggested various ways of implementing active traffic management strategies that interact directly with CAVs via messages exchanged among vehicles or with the infrastructure; however, due to the limitations above, such strategies have not been implemented. Efforts in this area should give the possibility to the main stakeholders involved in traffic management, namely, public agency operators and (toll) road operators, to exchange active traffic management-related messages with CAVs. Vehicle manufacturers and OEMs should consider such features in the design of future CAV systems aiming at a harmonized integration of CAVs within the transport system, also considering design principles that are not selfish but would lead to collective systemic benefits.

In the meantime, traffic management stakeholders, such as public agency operators and (toll) road operators, need to adapt their Traffic Management Center (TMC) policies and procedures to account for the increasing presence of CAVs so that both existing and novel traffic management measures, involving, for example, flow metering, variable speed limits, lane use management (including hard shoulder running), dynamic rerouting, and, in general, real-time messaging, are able to seamlessly interact with the operation of CAVs through the use of connected vehicle-to-infrastructure (V2I) communications. This will require an upgrade of the (mostly digital) infrastructure, to account for the improved sensing and actuating capabilities that will be moved from the infrastructure to the CAVs, as well as the development of data fusion engines, which would be capable of processing various types of traffic and vehicle data, resulting in efficient monitoring of the operational traffic situation.

Research efforts should support such developments and actively enable such transitions focusing, among other aspects, on the dynamic responses needed to be designed into the CAV, the standardization of these messages, responses to that messaging, as well as integration with the upgraded digital infrastructure and its novel decision support system widely employing innovative data fusion engines.

3 Conclusions and Future Needs

This chapter reflects and builds on some of the discussions that took place during the session on "Interactive traffic management for highly automated vehicles" at the Automated Road Transportation Symposium – ARTS 22.

It is becoming obvious that traffic management is an essential enabler for highly automated vehicles to be part of a safe and efficient traffic system, where a certain level of support from and interaction with the infrastructure are necessary requirements for future traffic systems with automated vehicles (e.g., for improving the ODD or implementing traffic rules). In addition, traffic management should serve to address unanticipated negative effects of automated vehicles (e.g., impacts on traffic efficiency), which are otherwise expected to decrease the current road and network capacities, with potentially negative economic externalities.

Finally, understanding and incorporating human factors in automated vehicles and traffic management design, regarding both strategies and responses, is essential in order to raise acceptability, ensure wider adoption of vehicle automation, and generate positive effects for society as a whole.

To achieve all of the above, there is a need to seek more collaboration among stakeholders, e.g., involving vehicle manufacturers, road authorities, and researchers to better frame the problems and achieve better, more sustainable, solutions through multidisciplinary approaches.

In particular, further research is needed on human factors, involving both the automated vehicle occupants and the interactions with other entities (e.g., interactions with human-driven vehicles in mixed traffic conditions, with pedestrians and other active modes, and multimodal traffic in general). Finally, the acceptability of traffic management measures, when not coded as rules, should be better investigated.

References

1. Geißler, T., Shi, E.: Taxonomies of connected cooperative and automated mobility. In: Proceedings of the 2022 IEEE Intelligent Vehicles Symposium, pp. 1517–1524 (2022). https://doi.org/10.1109/IV51971.2022.9827245
2. Khastgir, S., et al.: Report on ODD-ISAD architecture and NRA governance structure to ensure ODD compatibility. TM4CAD Deliverable D2.1. CEDR (2022). https://tm4cad.project.cedr.eu/deliverables/TM4CAD%20D2.1_submitted.pdf
3. Ntousakis, I.A., Nikolos, I.K., Papageorgiou, M.: On microscopic modelling of adaptive cruise control systems. Transp. Res. Procedia **6**, 111–127 (2015). https://doi.org/10.1016/j.trpro.2015.03.010
4. Mattas, K., et al.: Simulating Deployment of Connectivity and Automation on the Antwerp Ring Road. IET Intel. Transport Syst. **12**(9), 1036–1044 (2018). https://doi.org/10.1049/iet-its.2018.5287
5. Aittoniemi, E.: Evidence on Impacts of Automated Vehicles on Traffic Flow Efficiency and Emissions: Systematic Review. IET Intel. Transport Syst. **16**(10), 1306–1327 (2022). https://doi.org/10.1049/itr2.12219

6. Papamichail, I., et al.: Motorway traffic flow modelling, estimation and control with vehicle automation and communication systems. Ann. Rev. Control **48**, 325–46 (2019). https://doi.org/10.1016/j.arcontrol.2019.09.002
7. Karimi, M., Roncoli, C., Alecsandru, C., Papageorgiou, M.: Cooperative merging control via trajectory optimization in mixed vehicular traffic. Transp. Res. Part C: Emerg. Technol. **116**, 102663 (2020). https://doi.org/10.1016/j.trc.2020.102663
8. Sarker, A., et al.: A review of sensing and communication, human factors, and controller aspects for information-aware connected and automated vehicles. IEEE Trans. Intell. Transp. Syst. **21**(1), 7–29 (2020). https://doi.org/10.1109/TITS.2019.2892399
9. United Nations Economic and Social Council, New Assessment/Test Method for Automated Driving (NATM) Guidelines for Validating Automated Driving System (ADS), World Forum for Harmonization of Vehicle Regulations 187th session (2022). https://unece.org/sites/default/files/2022-04/ECE-TRANS-WP.29-2022-58.pdf
10. British Standards Institution (BSI). Operational design domain (ODD) taxonomy for an automated driving system (ADS), PAS 1883 (2020). https://standardsdevelopment.bsigroup.com/projects/2019-03092
11. Vreeswijk, J., Habibovic, A., Madland, O., Hooft, F.: Remote Support for Automated Vehicle Operations. In: Road Vehicle Automation 9, Springer, Cham (2022). https://doi.org/10.1007/978-3-031-11112-9_12
12. Bishop, R.: Intelligent Vehicle Technology and Trends, Artech House (2005)
13. Diakaki, C., Papageorgiou, M., Papamichail, I., Nikolos. I.: Overview and analysis of vehicle automation and communication systems from a motorway traffic management perspective. Transp. Res. Part A: Pol. Pract. **75**, 147–65, 2015. https://doi.org/10.1016/j.tra.2015.03.015
14. Message Sets for Advanced Traveler Information System (ATIS), SAE J2354_201906, in SAE Surface Vehicle Standard Standards (2019)

Author Index

G. Meyer and S. Beiker (Eds.): ARTSymposium 2022, LNMOB, p. 183, 2023.
https://doi.org/10.1007/978-3-031-34757-3